Success With Trigonometry

AUTHOR
John Prince
B.Math., A.Mus.
Mathematics Specialist
Lester B. Pearson High School
Burlington, Ontario, Canada

EDITOR
Jo Anne Robinson
B.Ed. (Mathematics), M.Ed. (Mathematics Curriculum)
Mathematics Specialist
President, *Washington State Mathematics Council*
No Limit Consultant
Everett, Washington, USA

SUCCESS WITH TRIGONOMETRY

Copyright © 2007 by John Prince

All Rights Reserved

ISBN 978-1-4357-0502-9

MY MATH TUTOR
CAME OVER YESTERDAY

Success With

Trigonometry

THERE ONCE WAS
A MAN
NAMED PYTHAGORAS

Pythagoras and others have stated: "The squares constructed on the sides of a 90°, or *right*, triangle have areas that are related. The area of the largest square has the same area as the other two squares combined."

I was asked once if "Theorem" was the last name of "Pythagoras". The answer is no.

Although this theorem appears to have existed before Pythagoras' time, the stated property continues to be important in mathematics and many trades.

The following triangle then, has the *area* formula: $x^2 + y^2 = r^2$.

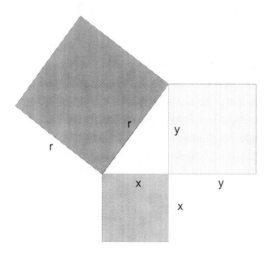

The longest side in a right triangle is called the *hypotenuse*, and is always opposite the 90° angle, the largest angle.

Call the longest side the 'hippopotamus', if that helps.

Since the other angles are smaller, the lengths of the side opposite these angles will also be shorter than the hypotenuse, and so, the areas of the squares formed on those sides will also be smaller.

The most common 90° triangle has the two smaller sides with lengths of 3 units, and 4 units. If the units are the same, the hypotenuse length can be determined using Pythagoras' area equation (Pythagoras' Theorem):

$$r^2 = 3^2 + 4^2$$
$$r^2 = 9 + 16$$
$$r^2 = 25$$
$$r = 5$$

Note: the variable 'r' has the exponent 2, implying a maximum of 2 answers. The other answer would be $r = -5$, but since we are only using lengths, which are never negative, this other answer is not used.

The measurement units used for both sides will be the same for the hypotenuse.

In the 'really old days', the units would possibly be 'cubits', the distance from the elbow to the furthest tip of the fingers.

This 90° triangle is called the '3, 4, 5 triangle', and is commonly used in many trades including: framing, deck construction, foundation construction, ceramic tile installation, cupboard door construction, etc. *As long as the units are consistent*, any triangle with sides in the ratio 3:4:5 is guaranteed to be right angled. (This means a triangle with side lengths 6,8, and10, or 30, 40, and 50 will also be right-angled.)

CONSTRUCTION INSTRUCTIONS FOR A 90° ANGLE

1. Select the location of one permanent, or fixed side.

2. Select the location of the required 90° corner.

3. Select the units to be used.

4. Select the largest lengths possible for the required application that are multiples of 3, 4, and 5. *For the foundation required on the next page, consider using 30 feet, 40 feet, and 50 feet (for the hypotenuse).*

5. On the fixed side, measure 40 units from the corner.

6. Mark this new point.

7. Using 2 tape measures, or strings, with lengths 30 feet, and 50 feet, measure the shorter distance from the corner.

8. Measure the longer distance from the second point.

9. Keeping the measures as straight as possible, move both lengths until the two measures meet. The point where the two measures meet will be the required point for the perpendicular side.

On a piece of paper, a protractor would be sufficient, but for a large application such as a foundation, or a deck, any protractor will be very imprecise.

On a ceramic tile job I completed in an 'L' shaped room, it was necessary to reverse the lengths making the fixed length 3 metres, and the measuring lengths 4, and 5 metres.

Move both lengths until
the 2 dots overlap.

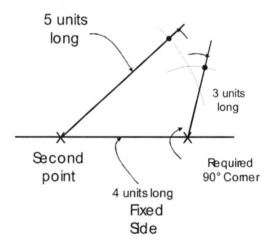

EXAMPLE 1

The size of any '3, 4, 5' triangle can be changed, if necessary, by multiplying each side by the same scale factor. Using a scale factor of two, all lengths will be doubled. This means a '6, 8, 10' triangle will also be right angled. Let's do a calculation check using the two shorter lengths.

$$r^2 = 6^2 + 8^2$$

$$r^2 = 36 + 64$$

$$r^2 = 100$$

$$r = 10$$

This means the '6, 8, 10' triangle is right-angled with the side length of 10 opposite the 90° angle.

By measuring with a protractor, you will notice none of the measures of the angles changed when this triangle is compared to the '3, 4, 5 triangle'. Check the measures of the angles in the diagrams on the next page. These triangles were constructed to be in the same ratio. Of course, the angle sum remains 180° for each triangle. The ratio of the lengths of any matching pair of sides is 2:1 when the larger '6, 8, 10' triangle is compared to the smaller '3, 4, 5' triangle.

Slope has been defined as a ratio, but written as a fraction.

Then $slope = \dfrac{rise}{run} = \dfrac{\Delta y}{\Delta x} = \dfrac{y_2 - y_1}{x_2 - x_1}$. None of these definitions involved an angle. A ratio does exist that involves an angle. This ratio is called *tangent*, or '*tan*' for short. It is defined:

$\tan B = slope = \dfrac{rise}{run} = \dfrac{\Delta y}{\Delta x} = \dfrac{y_2 - y_1}{x_2 - x_1}$.

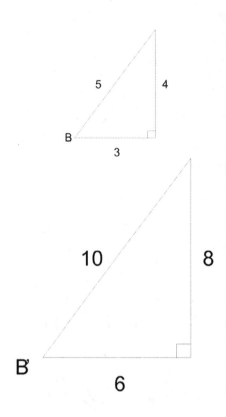

In the original triangle, the angle can be determined using the ratio equation

$$\tan B = \frac{4}{3}$$

For a 6, 8, 10 triangle, the ratio equation would be

$$\tan B' = \frac{8}{6} = \frac{4}{3}$$

The tan ratio is available on a calculator. In this example, using a protractor and exact drawings, the angles B, and B' measure approximately 53°. Thus can be checked by entering $\tan 53°$, where the answer will be approximately $\dfrac{4}{3} \approx 1.33$.

For a check of the angle, use one of the following:

$\boxed{\text{Tan}}$ 53.1 $\boxed{=}$, or

53.1 $\boxed{\text{tan}}$

Circle the method that works for your calculator. Use the same calculator each time.

To calculate the measure of the angle, without measuring or guessing, an equation must be solved. For the smaller triangle, solve $B = \tan^{-1}\left(\dfrac{4}{3}\right)$.

For the larger triangle, solve $B' = \tan^{-1}\left(\dfrac{8}{6}\right)$ Use the method on the right that works for the calculator you are using.

To calculate the measure of the angle, use one of the following sequences:

$\boxed{2^{nd}}$ $\boxed{\text{tan}}$ 4 $\boxed{\text{a b/c}}$ 3 $\boxed{=}$

$\boxed{2^{nd}}$ $\boxed{\text{tan}}$ $\boxed{(}$ 4 $\boxed{/}$ 3 $\boxed{)}$ $\boxed{=}$

4 $\boxed{\text{a b/c}}$ 3 $\boxed{2^{nd}}$ $\boxed{\text{tan}}$

4 $\boxed{/}$ 3 $\boxed{=}$ $\boxed{2^{nd}}$ $\boxed{\text{tan}}$

You need to set your calculator to degree mode. When it is set correctly, the display will show a smaller 'deg', or 'drg'. If the mode is different, use the 'mode' button, or 'deg' type buttons to change the mode. A check can be made by entering $\tan 45$ into the calculator. The answer is '1'.

On some calculators the $\boxed{2^{nd}}$ key is the $\boxed{\text{shift}}$ key.

Circle the method that works for your calculator. Use the same calculator each time.

EXAMPLE 2

Safety rules for the use of a ladder dictate its correct position: "one foot out for every four feet up". Basically, a slope of 4 is safe. Determine the angle generated by this ratio.

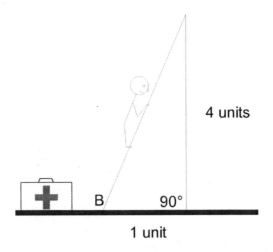

The angle at the foot of the ladder has been labeled B. (It is standard that upper case letters are usually angles, and sides are usually labeled in lower case letters. The opposite side to an angle usually has the same letter as the angle, but in lower case. In this case, the height of the triangle would be labeled as side 'b'.) Labeling the opposite side to an angle is easiest if you imagine yourself standing at the point, then look across at the opposite side.

The angle made by the ladder with the ground can be determined using the tangent ratio.

$$\tan B = slope$$
$$\tan B = \frac{4}{1}$$
$$B = \tan^{-1}(4)$$
$$B \approx 76.0°$$

EXAMPLE 3

Determine the measure of angle '*C*' in the following diagram.

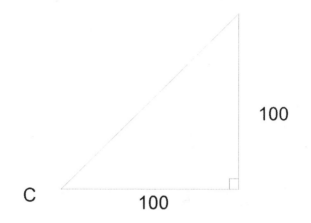

$$\tan C = slope$$

$$\tan C = \frac{100}{100}$$

$$\tan C = 1$$

$$C = \tan^{-1}(1)$$

$$C = 45°$$

The 45° angle is significant. In football, when the ball initially follows a 45° angle relative to the ground, the maximum horizontal distance will be possible for that play.

Gravity will reduce this angle continually until the ball hits the ground, or it is caught. So, for that final play of the game, launch the ball, as fast as possible at a 45° angle for maximum distance.

This answer for the measure of this angle is no surprise since the triangle has two equal side lengths. This makes the two angles opposite them, have the same measure. Since the three angles have a sum of 180°, the sum of the two equal angles will be 90° making each equal angle measure 45°.

This question is all yours.

QUESTION 4

Determine the measure of angle 'Q' in the diagram. The diagram is not drawn to scale.

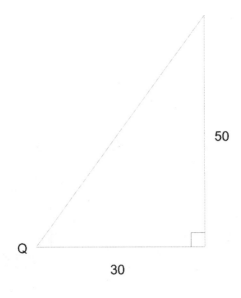

YOUR SOLUTION   *Measure of angle 'Q'*

QUESTION 4   POSSIBLE SOLUTION   *Measure of angle 'Q'*

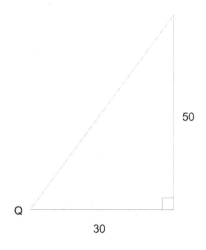

$$\tan Q = slope$$

$$\tan Q = \frac{50}{30}$$

$$Q = \tan^{-1}\left(\frac{50}{30}\right)$$

$$Q \approx 59.0°$$

The triangle is setup in the usual format: angle on the left, 90° on the right.

The measure of the angle is expected larger than 45° since, from the diagram, the height is longer than the base. (If height and base were equal, angle 'Q' would have measured 45°.)

One more to solve.

QUESTION 5

Determine the measure of angle 'R' in the diagram (not drawn to scale). Show a complete solution.

YOUR SOLUTION   *Measure of angle 'R'*

QUESTION 5   POSSIBLE SOLUTION   *Measure of angle 'R'*

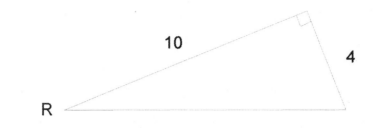

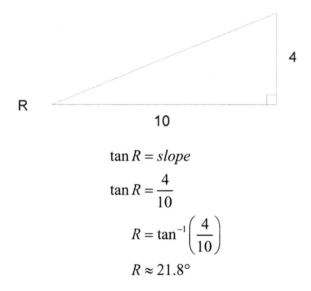

$$\tan R = slope$$

$$\tan R = \frac{4}{10}$$

$$R = \tan^{-1}\left(\frac{4}{10}\right)$$

$$R \approx 21.8°$$

Flip this triangle to show the slope.

Trace the triangle first if that helps.

Note: the tangent equation presented so far, requires slope to be positive.

(Sometimes it may be necessary to rotate and/or flip the triangle to show slope.)

The measure of the angle is less than 45° as expected, since the height is not as long as the base.

You Look Familiar

Last session, the tangent ratio was defined as equivalent to slope. The definition was given: $\tan B = slope = \dfrac{rise}{run} = \dfrac{\Delta y}{\Delta x} = \dfrac{y_2 - y_1}{x_2 - x_1}$. The corresponding diagram always had angle B on the left, and the 90° angle to the right of it.

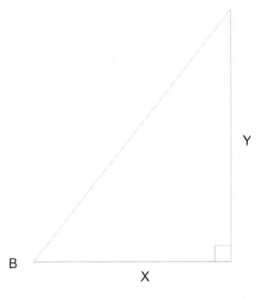

The angle at B was introduced since a triangle that was doubled in size had all side lengths doubled in size, but the corresponding, or matching, angles did not change. These triangles are called 'similar'.

Basically, the '3, 4, 5' and '6, 8, 10' triangles were 'Twins', but comparable to Arnold Schwarzenegger, and Danny deVito, rather than the Olsen's.

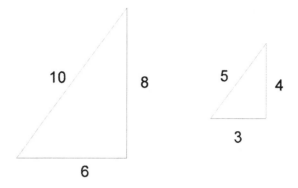

Although the first triangles had a 90° angle, any pair of triangles that have corresponding side lengths in the same ratio (in this case 2:1) are guaranteed to have corresponding angles equal. The reverse is also true. Corresponding equal angles generate similar triangles, and the corresponding sides are guaranteed to be in the same ratio.

Sometimes adjustments will be necessary to show the similar shapes. A reflection, or flip, may be required vertically, or horizontally. Possibly, a rotation may be required.

The two triangles above could have been arranged to have a common point as shown below. Still, it is recommended you separate the triangles to make them look similar.

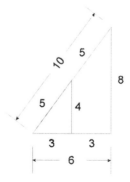

AREA

Since these triangles are right-angled, the heights are given. Using the area formula for a triangle, $A = \frac{1}{2}(base)(height)$, the area of each triangle can be calculated.

BIG TRIANGLE	SMALL TRIANGLE
$A = \frac{1}{2}(6)(8)$	$A = \frac{1}{2}(3)(4)$
$A = 24 \text{ units}^2$	$A = 6 \text{ units}^2$

The ratio of areas is 4:1 when the corresponding sides are in the ratio 2:1. *Both* the bass and the height were twice the size. This makes the larger area four times as large as the smaller area.

TREE HEIGHT WITHOUT A CALCULATOR

My parents needed a dead tree cut down in their backyard. The question was asked, "Where on the tree trunk should the cut be made so no damage would occur to another nearby tree when the dead tree was cut down?"

The tree was in the backyard so no cars were in any danger or being hit. All buildings were clear of any possible damage.

Ropes were used to direct the fall of the tree.

Minimal equipment is required.
1. a tape measure,
2. a level,
3. a piece of 8½ x 11 paper,
4. and duct tape, or rope.

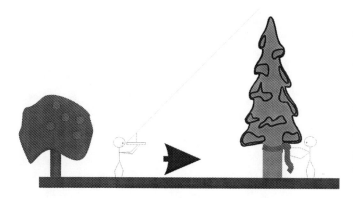

Be certain to move well away from the tree when it is cut.

The tree trunk can bounce a significant distance when it hits the ground, or the stump.

METHOD

1. Fold the paper over to generate a 45° angle at one of the corners.

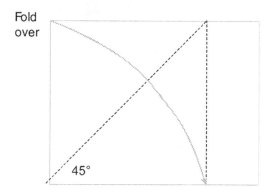

Fold over

45°

2. Hold the paper with the angle at one end of the level. Move this end so your eye uses the paper as the line of sight. (Use eye protection.)
3. Move towards, or away from the tree until the 45° angle lines up with the top of the tree, and the level is ... level. (Watch for obstacles.)
4. Have your buddy mark the tree where your level lines up on the trunk of the tree using the duct tape, or rope.
5. Your eye height, marked on the tree, predicts the location of the tree cut, so that the top of the tree falls where you are currently standing. (Use ropes to direct the fall of the tree.)

This example works since a similar triangle was generated. In this case, the horizontal and vertical sides had the same length. This guarantees the distance from the person measuring to the tree, is also the height of the cut portion of the tree. If the tree were cut off at ground level, an additional length equal to the height of your eyes, would need to be added to determine the tree height … unless the measurer was lying on the ground.

EXAMPLE 1

Determine the height of the hot air balloon.

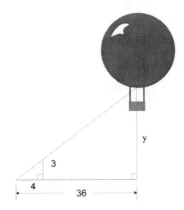

SOLUTION

1. First, separate the triangles.

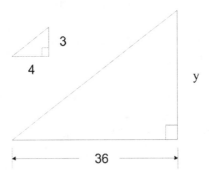

2. Organize the information. Usually a chart is ideal.

	Height	Base
Big Triangle		36
Small Triangle	3	4

3. Pick a letter of your choice for the balloon height. Earlier we used 'y' for vertical measurements.

	Height	Base
Big Triangle	y	36
Small Triangle	3	4

4. Generate the equation to solve, using ratios. You have two possibilities with the variable in the upper left position. $\dfrac{y}{3} = \dfrac{36}{4}$ or $\dfrac{y}{36} = \dfrac{3}{4}$. Both equations will have the same answer.

SOLUTION 1	SOLUTION 2

$$\frac{y}{3} = \frac{36}{4}$$

$$4y = 3(36)$$

$$4y = 108$$

$$\frac{4y}{4} = \frac{108}{4}$$

$$y = 27$$

$$\frac{y}{36} = \frac{3}{4}$$

$$4y = 3(36)$$

$$4y = 108$$

$$\frac{4y}{4} = \frac{108}{4}$$

$$y = 27$$

Special Case

Since both fractions are equal to each other, it is legal to generate another equation by cross multiplying.

This means the bottom of the balloon is 27 units off the ground. Since ratios are used, each separate triangle is required to have consistent units.

EXAMPLE 2

Determine the length shown as '*a*'.

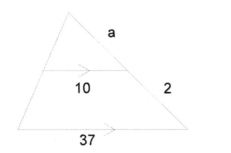

All units must be the same otherwise the problem cannot be solved. This will be more obvious when the triangles are separated.

SOLUTION

1. Separate the triangles.

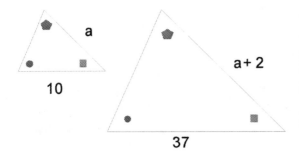

The top angles are equal since it is a shared angle. The other matching angles are equal since the bottom sides of the triangles are parallel.

2. Organize the information.

	Right Side	Bottom
Big Triangle	a+2	37
Small Triangle	a	10

3. Generate the equation to solve, using ratios. You have two possibilities. Consider the two ratio equations.

SOLUTION 1	SOLUTION 2	
$\dfrac{a+2}{a} = \dfrac{37}{10}$	$\dfrac{a+2}{37} = \dfrac{a}{10}$	*Cross-multiply.*
$10(a+2) = 37(a)$	$10(a+2) = 37(a)$	
$10a + 20 = 37a$	$10a + 20 = 37a$	*Again, the special case of*
$37a = 10a + 20$	$37a = 10a + 20$	
$37a - 10a = 20$	$37a - 10a = 20$	*one*
$27a = 20$	$27a = 20$	*fraction per side.*
$\dfrac{27a}{27} = \dfrac{20}{27}$	$\dfrac{27a}{27} = \dfrac{20}{27}$	
$a \approx 0.74$	$a \approx 0.74$	

EXAMPLE 3

Try this question on your own, but check the diagram of the separate triangles before starting the arithmetic.

Determine the length shown as 'c'.

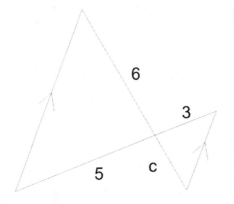

Move the triangles until they look similar.

Consider tracing them before redrawing them, and possibly rotating, or reflecting one or both triangles.

The units must be the same for the each separate triangle.

YOUR SOLUTION *Length of side 'c'*

EXAMPLE 3   POSSIBLE SOLUTION   *Length of side 'c'*

Separate the triangles.

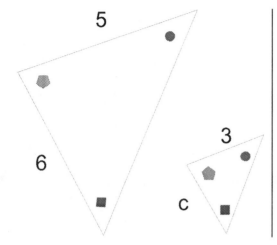

5

3

6

c

Remember, on the original diagram, the alternate angles will be equal since there are two parallel lines.

(Lower left angle = upper right angle and, upper left angle = lower right angle.)

Organize the information.

	Left Side	Top
Big Triangle	6	5
Small Triangle	c	3

Ratios and calculations:

SOLUTION 1	SOLUTION 2

$$\frac{6}{c} = \frac{5}{3}$$

$$5c = 6(3)$$

$$5c = 18$$

$$\frac{5c}{5} = \frac{18}{5}$$

$$c = 3.6$$

$$\frac{6}{5} = \frac{c}{3}$$

$$5c = 6(3)$$

$$5c = 18$$

$$\frac{5c}{5} = \frac{18}{5}$$

$$c = 3.6$$

Cross-multiply.

Again, the special case of one fraction per side.

QUESTION 4

Determine the lengths of the missing sides. All lengths have the same units.

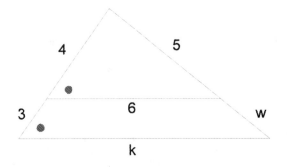

The two sides that look parallel have not been marked that way, but they will be parallel because of the equal angles shown.

YOUR SOLUTION   *Missing side lengths*

QUESTION 4   POSSIBLE SOLUTION   *Missing side lengths*

Separate the triangles.

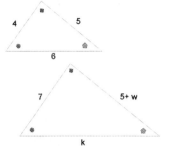

Although the bottom sides were not marked parallel, there are two equal angles (the top one was common to both triangles). This makes the 3rd angles also equal.

Organize the information.

	Left Side	Bottom	Right Side
Big Triangle	7	k	w+5
Small Triangle	4	6	5

Calculating the length of the "bottom" first …

SOLUTION 1	SOLUTION 2	
$\dfrac{7}{4}=\dfrac{k}{6}$	$\dfrac{7}{k}=\dfrac{4}{6}$	*Again, the special case of one fraction per side.*
$4k=7(6)$	$4k=7(6)$	
$4k=42$	$4k=42$	
$\dfrac{4k}{4}=\dfrac{42}{4}$	$\dfrac{4k}{4}=\dfrac{42}{4}$	*Cross-multiply.*
$k=10.5$	$k=10.5$	

Now calculate the length of the "right side".

SOLUTION 1	SOLUTION 2	SOLUTION 3	
$\dfrac{7}{w+5}=\dfrac{4}{5}$	$\dfrac{7}{4}=\dfrac{w+5}{5}$	$\dfrac{10.5}{6}=\dfrac{w+5}{5}$	*Again, the special case of one fraction per side.*
$4(w+5)=7(5)$	$4(w+5)=7(5)$	$6(w+5)=5(10.5)$	
$4w+20=35$	$4w+20=35$	$6w+30=52.5$	
$4w=35-20$	$4w=35-20$	$6w=52.5-30$	
$4w=15$	$4w=15$	$6w=22.5$	*Cross-multiply*
$\dfrac{4w}{4}=\dfrac{15}{4}$	$\dfrac{4w}{4}=\dfrac{15}{4}$	$\dfrac{6w}{6}=\dfrac{22.5}{6}$	
$w=3.75$	$w=3.75$	$w=3.75$	

QUESTION 5

Determine the length of the missing side. All lengths have the same units.

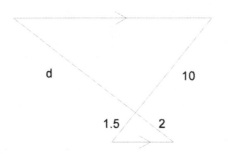

The diagram is not drawn to scale for lengths or angles.

YOUR SOLUTION   *Side length*

QUESTION 5 POSSIBLE SOLUTION *Side length*

Separate the triangles.

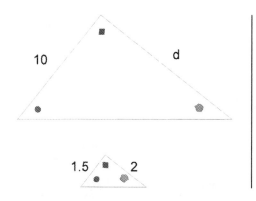

Be careful on this one.

The top triangle was rotated after the equal angles were marked.

Organize the information.

	Left Side	Right Side
Big Triangle	10	d
Small Triangle	1.5	2

Ratios and calculations:

SOLUTION 1	SOLUTION 2
$\dfrac{10}{d} = \dfrac{1.5}{2}$	$\dfrac{10}{1.5} = \dfrac{d}{2}$
$1.5d = 10(2)$	$1.5d = 10(2)$
$1.5d = 20$	$1.5d = 20$
$\dfrac{1.5d}{1.5} = \dfrac{20}{1.5}$	$\dfrac{1.5d}{1.5} = \dfrac{20}{1.5}$
$d \approx 13.3$	$d \approx 13.3$

Again, the special case of one fraction per side.

Cross-multiply.

THE
THREE
AMIGOS

In the first session slope was defined as a ratio but written as.

$$\tan B = slope = \frac{\Delta y}{\Delta x} = \frac{y_2 - y_1}{x_2 - x_1} = \frac{rise}{run}.$$ In each case the angle

was positioned on the left corner of a right-angle triangle.

There are two other basic ratios called *sine*, and *cosine*. On your calculator, the ratios are in sequence *sin*, *cos*, and *tan*. To make the definitions simpler, the sides will be defined as *hypotenuse* (opposite the 90° angle), *opposite* (the side opposite the angle you select), and *adjacent* (the only side left ... beside the selected angle.)

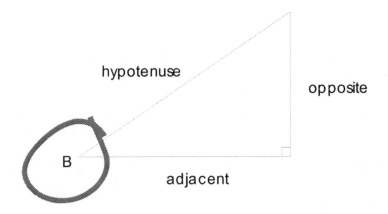

If the top angle was selected instead, the opposite and adjacent sides would also be switched.

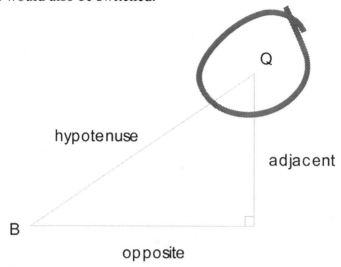

The 90° angle is not to be selected, since the hypotenuse is *always* the name given to this side, and it cannot have a second name as well.

The ratio definitions are:

$$\sin B = \frac{opposite}{hypotenuse}$$

$$\cos B = \frac{adjacent}{hypotenuse}$$

$$\tan B = "slope" = \frac{opposite}{adjacent}$$

The definitions can be arranged in a small 'tic-tac-toe' grid, using only the 1st letters of each name. The other two grids illustrate the same definitions. The grids will organize the triangle information, exposing the ratio to be used.

S	O	H
C	A	H
T	O	A

sine	opposite	hypotenuse
cosine	adjacent	hypotenuse
tangent	opposite	adjacent

sin	opp	hyp
cos	adj	hyp
tan	opp	adj

EXAMPLE 1

Determine the height of the tree as shown in this picture.

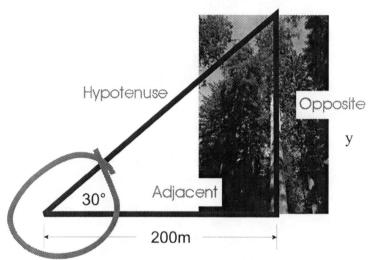

Courtesy of National Park Service.

Since the 30° angle was given, select it to work with. This makes the tree height the opposite side, and the bottom of the triangle, the adjacent side. The hypotenuse will always be opposite the 90° angle.

Using the grid definition from the previous page, circle the side you have, and the side you need. The line with 2 circles states the ratio equation to use. Write this equation, and solve it.

S O H
C A H
T O A

$$\tan 30° = \frac{opposite}{adjacent}$$

$$\tan 30° = \frac{y}{200}$$

$$\frac{\tan 30°}{1} = \frac{y}{200}$$

$$y = 200 \tan 30°$$

$$y \approx 115.47$$

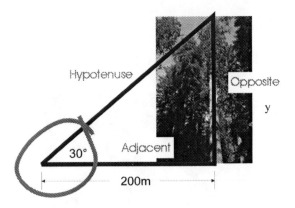

The height of the tree as shown is approximately 116m.

This height is only reasonable in the Redwood forest, California, where trees do grow to 120m (approx 472 feet). The trees regularly live for 600 years, and can reach ages of 2000 years. This website will provide additional information.
http://www.nps.gov/redw/

PROCEDURE TO SOLVE 90° TRIANGLES

1. Label the hypotenuse (opposite the 90° angle).
2. Select an angle to work with, excluding the 90° angle.
   (I recommend selecting the angle you were given.)
3. Label the side opposite this angle as the *opposite* side.
   (Imagine yourself standing at the angle location, then look across to the "opposite side".)
4. Label the remaining side as the *adjacent* side.
5. Write the grid definition in the tic-tac-toe format.
6. In the grid, circle the side you have the length of.
7. In the grid, circle the side you need to calculate the length of.
8. The horizontal line with 2 circles, states the ratio equation to use. Write this equation.
9. Solve the equation.
10. Check that the answer makes sense for this question. *(In this case, most trees would not grow this tall, but the height makes sense in the Redwood forest.)*

QUESTION 2

Cable supports are required for the following hot air balloon.
Determine the minimum length of one of these cables.

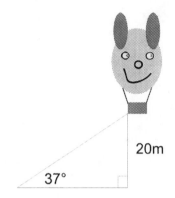

20m

37°

YOUR SOLUTION   *Cable length*

QUESTION 2   POSSIBLE SOLUTION   *Cable length*

1. Label the triangle.

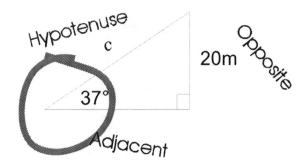

2. Determine which ratio to use, and then solve.

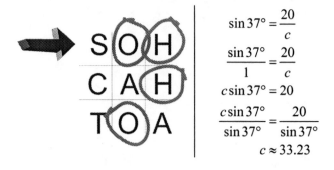

$$\sin 37° = \frac{20}{c}$$

$$\frac{\sin 37°}{1} = \frac{20}{c}$$

$$c \sin 37° = 20$$

$$\frac{c \sin 37°}{\sin 37°} = \frac{20}{\sin 37°}$$

$$c \approx 33.23$$

At least 33.2 metres of cable are required. Since the rounding did round 'off', the cable is definitely too short. Allow extra cable for connections etc.

QUESTION 3

Determine the height off the ground of the following kite. The length of string is 10m since the kite packaging includes this length as a maximum when the string has been fully extended. Assume the wind is blowing hard enough to make the string very close to a straight line, and the string is connected to the ground.

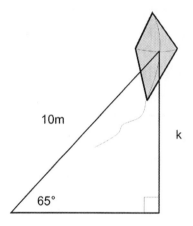

10m

k

65°

YOUR SOLUTION *Kite height*

QUESTION 3   POSSIBLE SOLUTION   *Kite height*

1. Label the triangle.

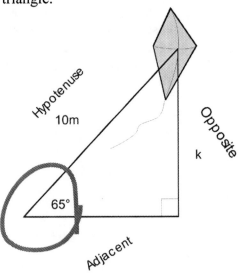

2. Determine which ratio to use, and then solve.

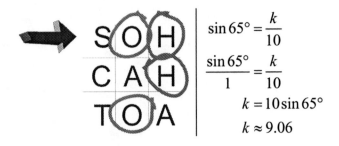

$$\sin 65° = \frac{k}{10}$$

$$\frac{\sin 65°}{1} = \frac{k}{10}$$

$$k = 10 \sin 65°$$

$$k \approx 9.06$$

The kite is approximately 9 metres off the ground. (Stay away from any hydro wires.)

QUESTION 4

A driving accident has occurred where a car has been driven
into the lake. The average angle of depression of the lake
bottom is known to be 18°. Based on the posted speed limit,
officers on the scene have calculated the distance the car is
likely to have traveled into the lake, to be 50 metres. Determine
the horizontal distance the car is from the shore. This distance
will be the distance a rescue boat will travel before dropping a
rescue cable and divers.

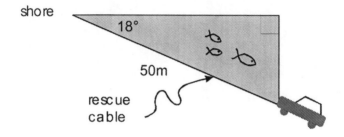

YOUR SOLUTION   *Horizontal distance*

QUESTION 4   POSSIBLE SOLUTION
Horizontal distance

1. Label the triangle.

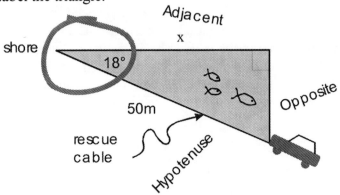

2. Determine which ratio to use, and then solve.

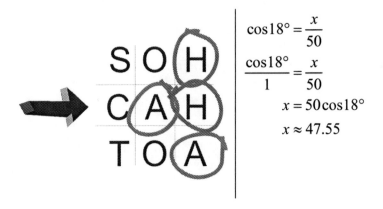

$$\cos 18° = \frac{x}{50}$$

$$\frac{\cos 18°}{1} = \frac{x}{50}$$

$$x = 50 \cos 18°$$

$$x \approx 47.55$$

The car is currently about 48 metres off shore. A rescue by tow truck or by any vehicle with a winch is possible since the car is reasonably close to shore.

QUESTION 5

Determine the horizontal distance between the following buildings. The angle of depression has been measured by a "protractor", or equivalent, to be 54°, as shown.

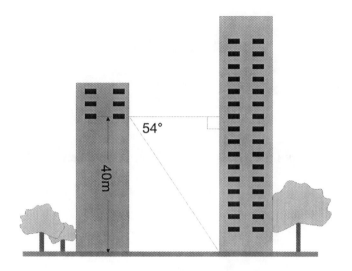

YOUR SOLUTION *Distance between buildings*

QUESTION 5   POSSIBLE SOLUTION
Distance between buildings

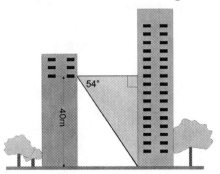

There are two obvious triangles. The "safer" solution is to use
the triangle with the most given information. I have selected the
triangle as shown.

1. Label the triangle.

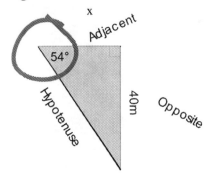

2. Determine which ratio to use, and then solve.

$$\tan 54° = \frac{40}{x}$$

$$\frac{\tan 54°}{1} = \frac{40}{x}$$

$$x(\tan 54°) = 40$$

$$\frac{x(\tan 54°)}{\tan 54°} = \frac{40}{\tan 54°}$$

$$x = \frac{40}{\tan 54°}$$

$$x \approx 29.06$$

So, the distance between the buildings is approximately 29 m.

ALTERNATE SOLUTION: *Distance between buildings*

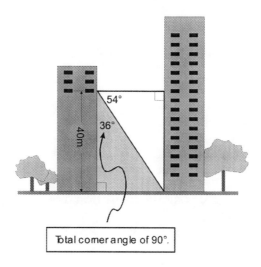

Total corner angle of 90°.

1. Label the triangle.

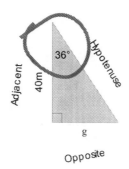

2. Determine which ratio to use, and then solve.

S O H $\tan 36° = \dfrac{g}{40}$

C A H $\dfrac{\tan 36°}{1} = \dfrac{g}{40}$

T O A $g = 40 \tan 36°$

$g \approx 29.06$

Again, the distance between the buildings is approximately 29m.

WHAT'S
MY
ANGLE ?

In the first session, the slope ratio was defined as

$$slope = m = \frac{\Delta y}{\Delta x} = \frac{y_2 - y_1}{x_2 - x_1} = \tan B$$

with B defined as the angle positioned left from the 90° angle, and a positive slope required.

To calculate the measure of the angle, when the opposite and adjacent sides were given, the $\tan^{-1}$ key was required. (On most calculators, this key is accessed when the 'SHIFT', or 2nd key is pressed, followed by the 'tan' key.)

The first example used the common 3, 4, 5 triangle. Initially, we showed the hypotenuse length to be 5, and then we solved the equation.

$$B = \tan^{-1}\left(\frac{4}{3}\right).$$

$$B \approx 53.1°$$

This angle of 53.1°, between the hypotenuse and the base of the triangle, is the same as a slope of $\frac{4}{3}$ when the left corner of the triangle is positioned at the point $(0,0)$, and the base of the triangle rests on the x axis.

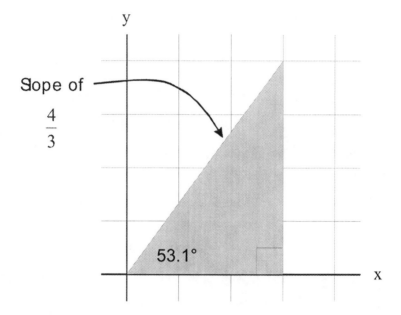

Since we now know the measure of all three sides, *any pair* of sides can be used with their matching ratio: 'sin', 'cos', or 'tan'. The resulting angle should be the same each time. Let's try it.

Using the hypotenuse, and opposite side:

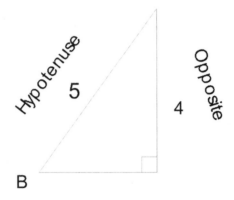

Using the tic-tac-toe definitions from last session, the following would occur.

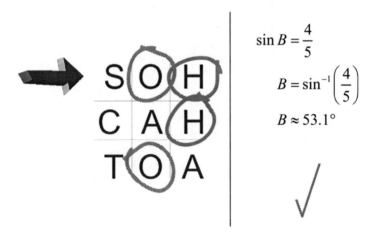

$$\sin B = \frac{4}{5}$$

$$B = \sin^{-1}\left(\frac{4}{5}\right)$$

$$B \approx 53.1°$$

Here is the final pair of sides.

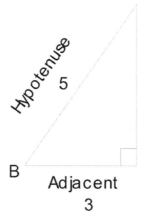

Again, using the tic-tac-toe definitions from last time, the following would occur.

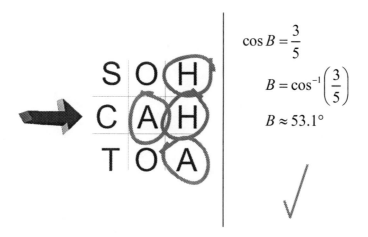

$$\cos B = \frac{3}{5}$$

$$B = \cos^{-1}\left(\frac{3}{5}\right)$$

$$B \approx 53.1°$$

The third angle can be calculated since the sum of the three angles is 180°, and we have two of them. In this case, the other angle will measure approximately 36.9°. (This angle could be given a name, and then solved using any of 'sin', 'cos', or 'tan', although the labeling of opposite, and adjacent would change.)

SUMMARY

1. When two sides of a right-angled triangle are given, any angle measure can be calculated.

2. On the calculator, the $\sin^{-1}, \cos^{-1}$, and $\tan^{-1}$ are required for calculating the measure of an angle.

3. The third side of a right-angled triangle can be determined using Pythagoras' Theorem.

QUESTION 1

Determine the measure of angle '*A*'.

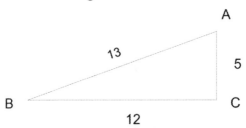

YOUR SOLUTION   *Angle 'A'*

QUESTION 1   POSSIBLE SOLUTION   *Angle 'A'*

The diagram is not drawn to scale, and is not guaranteed to be right-angled. Before any ratios, or side labeling, can occur, a right angle is required.

Since the longest side is 13 units long, the 90° angle could only be opposite this side. As well, angle 'A' will be the next biggest angle, since the side opposite to it, is the next longest.

Using Pythagoras Theorem for areas, we have the equation $5^2 + 12^2 = 13^2$ to check. Since the left side and right side of this equation both equal 169, the triangle has a 90° angle, and the three trig ratios can now be used successfully.

If the slope of the hypotenuse is kept positive, the position of the diagram would be adjusted to have the angle in the bottom left, but this is not a necessary step.

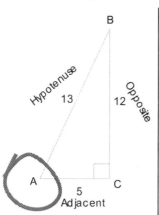

Since any pair of sides can be used, you can choose which pair you want.

All 3 equations produce the same answer.

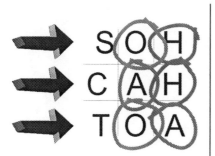

$$\sin A = \frac{12}{13} \qquad A = \sin^{-1}\left(\frac{12}{13}\right)$$

$$\cos A = \frac{5}{13} \qquad A = \cos^{-1}\left(\frac{5}{13}\right)$$

$$\tan A = \frac{12}{5} \qquad A = \tan^{-1}\left(\frac{12}{5}\right)$$

$$A \approx 67.4°$$

Question 2

A clock pendulum swings back and forth continuously through a specific distance. The angle of its swing can be calculated, but its not as exciting as the Xtreme Skyflyer™ at Canada's Wonderland, or the Ripcord at Cedar Point in Ohio. Both rides are 6 feet off the ground at their lowest point. Estimates of lengths have been made for the Ripcord ride as shown below. Determine the approximate angle the ride will pass through during the initial drop.

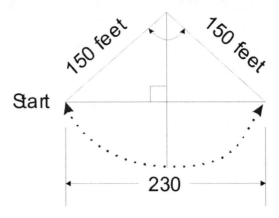

Your Solution   *Angle of travel*

QUESTION 2   POSSIBLE SOLUTION   *Angle of travel*

This time a 90° angle has been specified so, a Pythagoras check is not required.

Although the ride will not continue swinging back and forth, initially assume both smaller triangles measure the same. This makes the bottom of each triangle 115 feet. (The final calculated angle will be larger than the actual answer since gravity will "slow" the swing throughout the ride.)

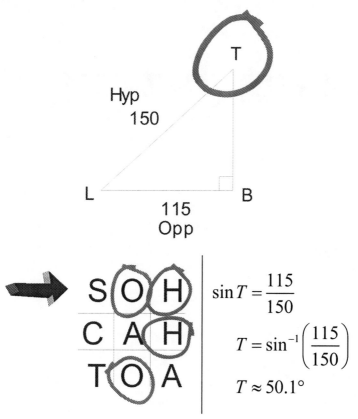

$$\sin T = \frac{115}{150}$$

$$T = \sin^{-1}\left(\frac{115}{150}\right)$$

$$T \approx 50.1°$$

On the first drop, the ride (and riders) will travel through approximately $50° \times 2 \approx 100°$.

QUESTION 3

To make a semicircular deck, the specifications for a piece of "wood" used are shown below. According to the instructions, a 3° cut is required to the original rectangular piece. Using the diagram, determine whether or not the number of degrees is correct.

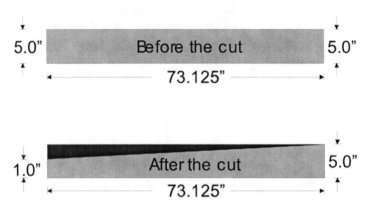

5.0"   Before the cut   5.0"

73.125"

1.0"   After the cut   5.0"

73.125"

YOUR SOLUTION   *Angle to cut*

QUESTION 3   POSSIBLE SOLUTION   *Angle to cut*

Contractors will have different tools available to make this cut, although it is possible to make the cut with a circular saw. A chalk line will be made to mark the required cut first.

The original 5.0″ end is to be tapered to 1.0″. There will be 4.0″ removed.

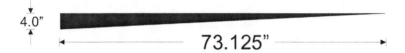

4.0″   73.125″

The measure of the smallest angle is to be calculated. This time the angle was left in its original position without turning the diagram.

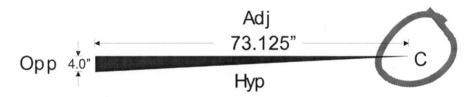

Adj
73.125″
Opp 4.0″   C
Hyp

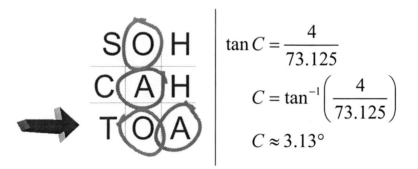

$$\tan C = \frac{4}{73.125}$$

$$C = \tan^{-1}\left(\frac{4}{73.125}\right)$$

$$C \approx 3.13°$$

The 3° angle is accurate. Although some equipment will be capable of cutting 3.13°, the 3° angle will be sufficient since the deck will have a space between each piece.

WHOSE EQUATION IS IT ANYWAYS ?

CALCULATOR REVIEW

Calculate the answer(s) to the following equations. Side lengths are to be accurate to 2 decimal places, and angle measurements are to be accurate to 1 decimal place.

$\sin 10° = \dfrac{a}{5}$	
$\cos B = 0.938$	
$\tan 30° = \dfrac{c}{45}$	
$\sin D = \dfrac{13}{20}$	
$\cos 50° = \dfrac{e}{30}$	
$\sin F = \dfrac{1.92}{3.4}$	
USE LOW GEAR 6 % GRADE FOR 2 km $\tan G = 6\%$	
$\cos 80° = \dfrac{10}{h}$	

CALCULATOR SOLUTIONS

$\sin 10° = \dfrac{a}{5}$	$\dfrac{\sin 10°}{1} = \dfrac{a}{5}$ $a = 5\sin 10°$ $a \approx 0.87$
$\cos B = 0.938$	$B = \cos^{-1}(0.938)$ $B \approx 20.3°$
$\tan 30° = \dfrac{c}{45}$	$\dfrac{\tan 30°}{1} = \dfrac{y}{45}$ $y = 45\tan 30°$ $y \approx 25.98$
$\sin D = \dfrac{13}{20}$	$D = \sin^{-1}\left(\dfrac{13}{20}\right)$ $D \approx 40.5°$
$\cos 50° = \dfrac{e}{30}$	$\dfrac{\cos 50°}{1} = \dfrac{e}{30}$ $e = 30\cos 50°$ $e \approx 19.28$
$\sin F = \dfrac{1.92}{3.4}$	$F = \sin^{-1}\left(\dfrac{1.92}{3.4}\right)$ $F \approx 34.4°$
USE LOW GEAR 6% GRADE FOR 2 km $\tan G = 6\%$	$\tan G = 0.06$ $G = \tan^{-1}(0.06)$ $G \approx 3.4°$
$\cos 80° = \dfrac{10}{h}$	$\dfrac{\cos 80°}{1} = \dfrac{10}{h}$ $h\cos 80° = 1(10)$ $\dfrac{h\cos 80°}{\cos 80°} = \dfrac{10}{\cos 80°}$ $h \approx 57.59$

EQUATION REVIEW

State the best equation for the following situations. Remember to label the angle and its opposite side with the same letters. Upper case for the angle, lower case for the side.

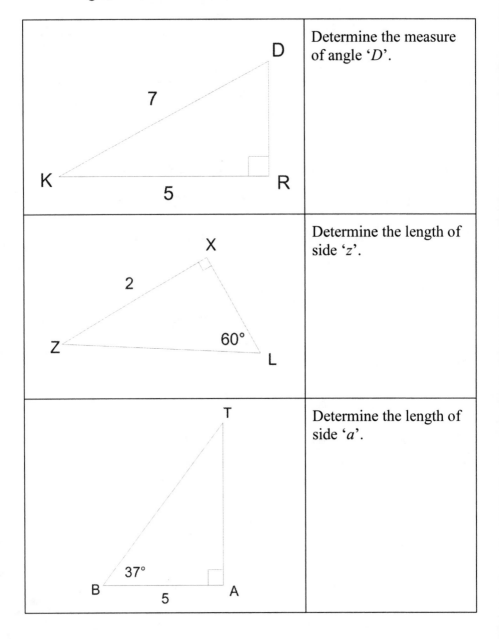

(triangle with D, 7, K, 5, R)	Determine the measure of angle 'D'.
(triangle with X, 2, Z, 60°, L)	Determine the length of side 'z'.
(triangle with T, 37°, B, 5, A)	Determine the length of side 'a'.

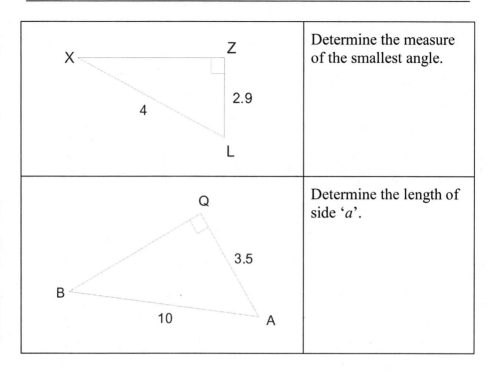

| | Determine the measure of the smallest angle. |

| | Determine the length of side '*a*'. |

EQUATION SOLUTIONS

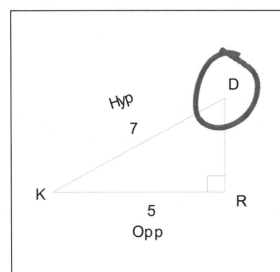

Measure of angle '*D*'.

$$\sin D = \frac{5}{7}$$

$$D = \sin^{-1}\left(\frac{5}{7}\right)$$

$$D \approx 45.6°$$

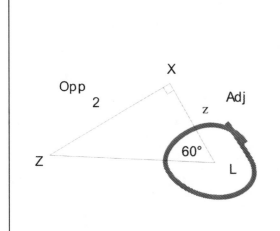

Length of side '*z*'.

$$\tan 60° = \frac{2}{z}$$

$$\frac{\tan 60°}{1} = \frac{2}{z}$$

$$z \tan 60° = 2$$

$$\frac{z \tan 60°}{\tan 60°} = \frac{2}{\tan 60°}$$

$$z \approx 1.15$$

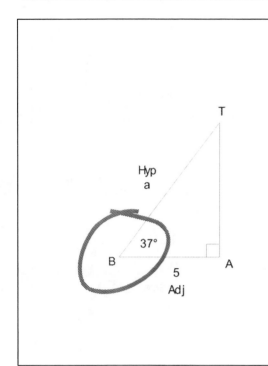

Length of side '*a*'.

$$\cos 37° = \frac{5}{a}$$

$$\frac{\cos 37°}{1} = \frac{5}{a}$$

$$a \cos 37° = 5$$

$$\frac{a \cos 37°}{\cos 37°} = \frac{5}{\cos 37°}$$

$$a \approx 6.26$$

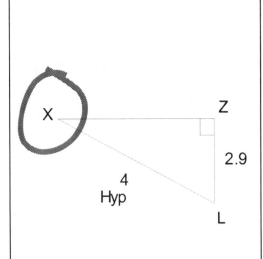

Measure of the smallest angle.

The smallest angle is opposite the shortest side

There are two possible solutions but the faster is to calculate the measure of angle X directly. If you think you do not have the smallest angle, calculate the other with the triangle angles totaling 180°.

The sketch is not drawn to scale.

$$\sin X = \frac{2.9}{4}$$

$$X = \sin^{-1}\left(\frac{2.9}{4}\right)$$

$$X \approx 46.5°$$

The smallest angle measure will be 90° - 46.5° = 43.5°. This is the measure of angle 'L'.

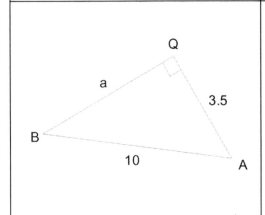

Length of side '*a*'.

The length of side 'a' is available using Pythagoras Theorem without using the trig ratios.

$$a^2 + 3.5^2 = 10^2$$

$$a^2 + 12.25 = 100$$

$$a^2 = 100 - 12.25$$

$$a^2 = 87.75$$

$$a = \sqrt{87.75}$$

$$a \approx 9.37$$

DIAGRAM CONSTRUCTION

The following equations were generated from a triangle, using some side(s), and some angle(s). Construct a triangle to illustrate a possible source for the equation. The triangles do not need to be constructed to scale.

$\tan B = \dfrac{4}{3}$	
$\sin 51° = \dfrac{y}{5}$	
$\cos C = 0.379$	
$\tan 28° = \dfrac{7}{x}$	
$x^2 = 10^2 - 3^2$	

DIAGRAM SOLUTIONS

The following equations were generated from a triangle, using some side(s), and some angle(s). Construct a triangle to illustrate a possible source for the equation. The triangles do not need to be constructed to scale.

$$\tan B = \frac{4}{3} = \frac{opp}{adj}$$	*This is a ratio equation with no guaranteed side lengths, however, these ratios give a length of 5 to the hypotenuse when the other lengths are 3, and 4 units.*
$$\sin 51° = \frac{y}{5}$$	*This time, the length of the hypotenuse is guaranteed to be 5 units.*

$\cos C = 0.379$	*This is a ratio equation with no guaranteed side lengths. The decimal can be written as the fraction* $\dfrac{0.379}{1}$. C — 1 — 0.379 (triangle)
$\tan 28° = \dfrac{7}{x}$	*The length of the opposite side is guaranteed to be 7 units.* 28°, 7, x (triangle)
$x^2 = 10^2 - 3^2$	*Lengths are guaranteed with a possible rewrite of the equation to* $x^2 + 3^2 = 10^2$. *The hypotenuse length will be 10 units.* 10, 3, x (triangle)

Applications

There are two angles that have specific positions relative to horizontal. If the angle is above horizontal, it is called an *angle of elevation*. If the angle is below horizontal, it is called an *angle of depression*.

Back in the session *The Three Amigos,* on page 31, the angle from the ground to the top of the tree was an *angle of elevation*.

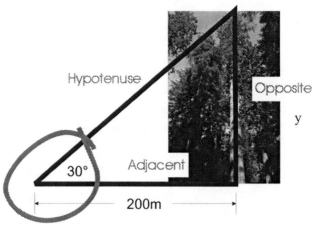

Courtesy of National Park Service.

The example with the car in the water, on page 38, generated an *angle of depression*.

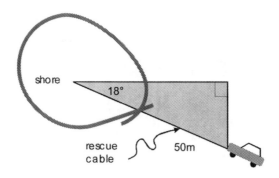

THAT'S NOT RIGHT

REVIEW

The formula $A = \dfrac{1}{2}bh$ is the most common formula to determine the area of a triangle. Determine the area of the following triangle, which as usual, is not guaranteed drawn to scale.

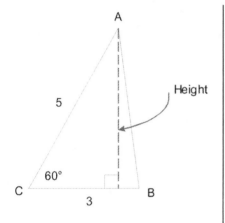

From previous work, you may be tempted to say the height is 4 units since the famous 3, 4, 5 triangle is right-angled. However, the angle of 60° at C is not the angle that occurs in the 3, 4, 5 triangle. It was 53.1°. (See page 7.) Recall from then:

$$\tan C = \frac{4}{3}$$

$$C = \tan^{-1}\left(\frac{4}{3}\right)$$

$$C \approx 53.1°$$

Since the angle here, 60°, is larger than the 53.1°, the height of the triangle is expected to be larger than 4 units.

So, let's calculate the height of the triangle using the given 60° angle, and the hypotenuse. The bottom length of the right triangle is not available, but its certainly not exactly 3 units.

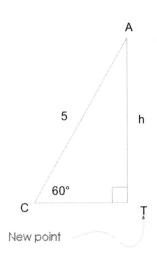

Pythagoras' Theorem is not currently useful since only one side length was guaranteed and the Theorem requires 2 sides be available.

$$\sin 60° = \frac{h}{5}$$

$$h = 5\sin 60°$$

$$h \approx 4.33$$

Although only 4.33 was recorded, keep all the decimals until the last calculation is done.

Now, this allows us to calculate the area as required.

$$Area = \frac{1}{2}(base)(height)$$

$$\approx \frac{1}{2}(3)(4.33)$$

$$\approx 6.50\ units^2$$

One method of keeping all the decimals uses the memory available on the calculator. The other, use the ANS (Answer) key that may be on your calculator.

This method was too much work. Certainly there are software packages that will generate an excellent approximation, but if we "massage" the equations just used, another area formula can be developed quickly.

List the two sections of the calculations together and then combine the equations by substitution. The full theory version is available in most textbooks.

$$Area = \frac{1}{2}(base)(height)$$

$$\approx \frac{1}{2}(3)(4.33)$$

$$\approx 6.50 \ units^2$$

$$\sin 60° = \frac{height}{5}$$

$$height = 5\sin 60°$$

$$height \approx 4.33$$

$$Area = \frac{1}{2}(base)(height)$$

$$= \frac{1}{2}(3)(4.3)$$

$$= 6.50 \ units^2$$

$$\sin 60° = \frac{height}{5}$$

$$height = 5\sin 60°$$

$$height = 4.3$$

$$Area = \frac{1}{2}(base)(height)$$

$$Area = \frac{1}{2}(3)(5\sin60°)$$

$$= \frac{1}{2}(3)(4.3)$$

$$= 6.50 \ units^2$$

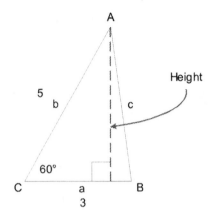

Matching the letters, we have:

$$Area = \frac{1}{2}(3)(5)(\sin 60°)$$

With the fully labeled diagram on the left, the equation becomes

$$Area = \frac{1}{2}ab\sin C$$

For a triangle labeled with each side the lower case of its opposite angle, the following can occur.

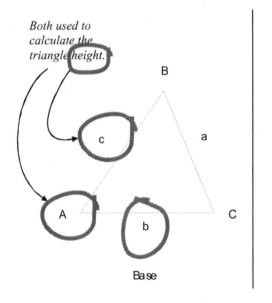

Both used to calculate the triangle height.

Base

$$Area = \frac{1}{2}(base)(height)$$

$$Area = \frac{1}{2}(b)(c\sin A)$$

$$Area = \frac{1}{2}bc\sin A$$

This new formula works any time there are 2 given side lengths and the measure of the angle between them is available.

Using the same method of calculating triangle height, other time-saving formulas are available. For example, determine the length of side 'c' from the same triangle.

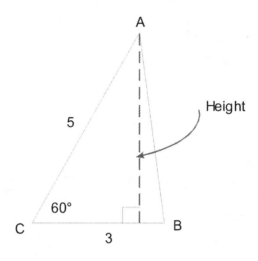

The height was calculated earlier.

$$\sin 60° = \frac{h}{5}$$

$$h = 5\sin 60°$$

$$h \approx 4.33$$

Side 'a' was not used since it is not a side of either 90° triangle.

Using another ratio, or using Pythagoras' Theorem, the lengths of the two sections of side 'a' can be determined. Separate side 'a' into two parts called x, and q, with $x + q = a$.

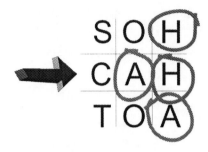

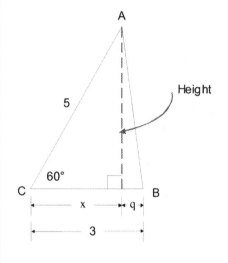

$$\cos 60° = \frac{x}{5}$$

$$x = 5\cos 60°$$

$$x = 2.50$$

By Pythagoras' Theorem

$$x^2 + h^2 = 5^2$$

$$x^2 + \left(5\sin 60°\right)^2 = 25$$

$$x^2 + 18.75 = 25$$

$$x^2 = 6.25$$

$$x = \sqrt{6.25}$$

$$x = 2.50$$

Since length is positive, the negative answer to the equation will not be used.

This makes the length 'q' = 0.5 (subtracting the length of side 'x' = 2.5 units, from side 'a' = 3).

Finally, the length of side 'c' is available using Pythagoras' Theorem only.

$$c^2 = q^2 + h^2$$

$$c^2 = \left(0.5\right)^2 + \left(5\sin 60°\right)^2$$

$$c^2 = 0.25 + 18.75$$

$$c^2 = 19$$

$$c = \sqrt{19}$$

$$c = 4.36$$

Now, this was a four step problem that involved the cosine ratio and Pythagoras' Theorem. There is a shorter (and better) way.

THE COSINE LAW

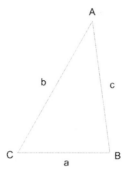

There are three available equations for this triangle. Each begins in a Pythagorean form, but each has a correction term. Most students just memorize the formula, but it is developed in Appendix A.

$$c^2 = a^2 + b^2 - 2ab(\cos C)$$
$$b^2 = a^2 + c^2 - 2ac(\cos B)$$
$$a^2 = b^2 + c^2 - 2bc(\cos A)$$

Note the relationships:

When the length of side 'c' is required the measure of angle 'C' is used.
When the length of side 'b' is required the measure of angle 'B' is used.
When the length of side 'a' is required the measure of angle 'A' is used.

In each case, if the angle used is 90°, the equation will simplify to a Pythagorean format since $\cos 90° = 0$.

As usual, any new formula is expected to work on a known example. Test one of these equations on the triangle we just used.

Determine the length of side '*c*' in this triangle.

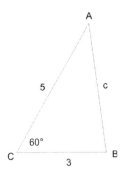

The length is $\sqrt{19} \approx 4.36$ units from the previous calculations

The original tic-tac-toe format with "opposite, adjacent, and hypotenuse" is not available since this triangle has no guaranteed 90° angle. List the angles and sides instead in a grid, and then use the circled information to select the correct equation.

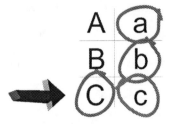

Since one line has two circled items, the grid suggests the equation should be $c^2 = a^2 + b^2 - 2ab(\cos C)$. Angle 'C', and its opposite side '*c*' are both circled. Using the triangle information, the equation becomes:

$$c^2 = a^2 + b^2 - 2ab(\cos C)$$
$$c^2 = (3)^2 + (5)^2 - 2(3)(5)\cos 60°$$
$$c^2 = 9 + 25 - 30\cos 60°$$
$$c^2 = 34 - 30(0.5)$$
$$c^2 = 34 - 15$$
$$c^2 = 19$$
$$c \approx 4.36$$

Watch the order of operations.

However, some calculators will complete all of these calculations correctly once the equation is entered.

The new formula gives the same answer as before, but requires fewer steps.

QUESTION 2

Determine the length of side '*q*'.

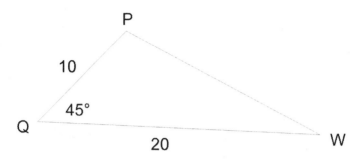

YOUR SOLUTION   *Side 'q'*

QUESTION 2   POSSIBLE SOLUTION   *Side 'q'*

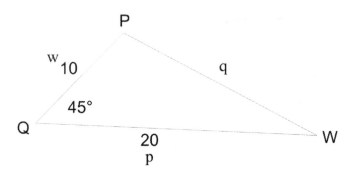

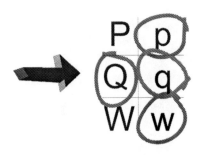

$$q^2 = p^2 + w^2 - 2pq(\cos Q)$$
$$q^2 = (20)^2 + (10)^2 - 2(20)(10)\cos 45°$$
$$q^2 = 400 + 100 - 400\cos 45°$$
$$q^2 \approx 500 - 400(0.7071)$$
$$q^2 \approx 500 - 282.84$$
$$q^2 = 217.16$$
$$q = \sqrt{217.16}$$
$$q \approx 14.74$$

QUESTION 3

Determine the length of side 'g'.

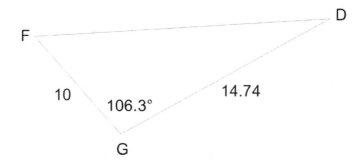

YOUR SOLUTION *Side 'g'*

QUESTION 3   POSSIBLE SOLUTION   *Side 'g'*

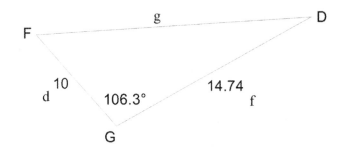

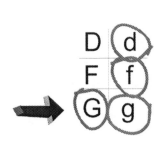

$$g^2 = d^2 + f^2 - 2df(\cos G)$$

$$g^2 = (10)^2 + (14.74)^2 - 2(10)(14.74)\cos106.3°$$

$$g^2 = 100 + 217.2676 - 294.8\cos106.3°$$

$$g^2 \approx 317.2676 - 294.8(-0.2807)$$

$$g^2 \approx 317.2676 + 82.7405$$

$$g^2 \approx 400.00$$

$$g = \sqrt{400}$$

$$g = 20.00$$

In this question, since angle G is the largest angle, side 'g' will be the longest side. The answer of 20 is consistent with this observation.

If a calculation error had occurred, usually it will be a sign error, and will have occurred as shown below.

$$g^2 \approx 317.2676 - 82.7405$$

$$g^2 \approx 234.53$$

$$g \approx \sqrt{234.53}$$

$$g \approx 15.31$$

Sometimes this side length will be clearly inconsistent with the diagram, but this time the length of side 'g' is still the longest. However, this would suggest the measure of angle 'F' is about the same as the measure of angle 'G'. With these 2 angles, the sum of the angles in the triangle is already over 180°.

QUESTION 4

Determine the length of side '*n*'.

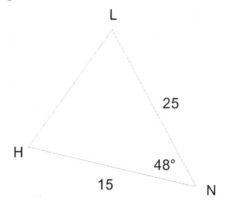

YOUR SOLUTION   *Side 'n'*

QUESTION 4   POSSIBLE SOLUTION   *Side 'n'*

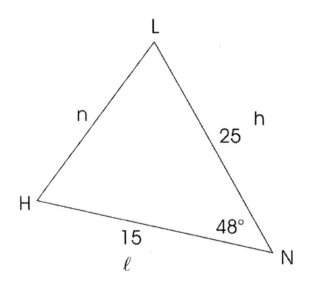

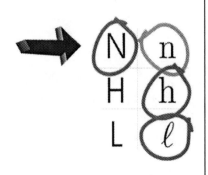

$$n^2 = h^2 + l^2 - 2hl\left(\cos N\right)$$

$$n^2 = \left(25\right)^2 + \left(15\right)^2 - 2\left(25\right)\left(15\right)\cos 48°$$

$$n^2 = 625 + 225 - 750\cos 48°$$

$$n^2 = 850 - 750\cos 48°$$

$$n^2 \approx 348.15$$

$$n^2 = \sqrt{348.15}$$

$$n \approx 18.66$$

THE ANGLE
AND
THE THREE SIDES

REVIEW

Determine the length of the missing side.

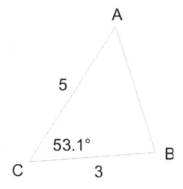

YOUR SOLUTION   *Side 'c'*

POSSIBLE SOLUTION   *Side 'c'*

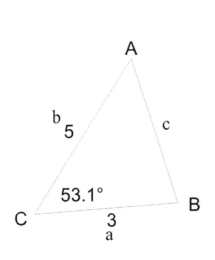

The triangle is not given as right-angled, limiting the best solution to the cosine law.

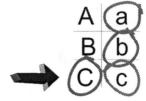

$$c^2 = a^2 + b^2 - 2ab(\cos C)$$

$$c^2 = (3)^2 + (5)^2 - 2(3)(5)\cos 53.1°$$

$$c^2 = 9 + 25 - 30\cos 53.1°$$

$$c^2 \approx 34 - 30(0.6004)$$

$$c^2 \approx 34 - 18.0126$$

$$c^2 = 15.9874$$

$$c = \sqrt{15.9874}$$

$$c \approx 4.00$$

If you remembered calculating the angle 53.1° back in the first session (see page 7), you would have recognized the triangle to be the '3, 4, 5' triangle. Notice though, you have just demonstrated the cosine law also works on a right-angled triangle. In fact, the cosine law is the most powerful of the trigonometric methods. However, it does require two or more sides, and most times, an angle contained, caught, or included between a pair of sides.

NEW

Determine the measure of the smallest angle in the following triangle.

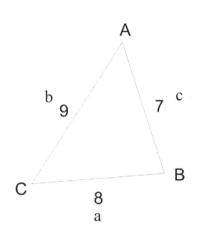

The smallest angle is opposite the shortest side.

The largest angle is opposite the longest side.

The middle sized angle is opposite the middle-sized side.

If you forget the pattern, you can always check the sizes by calculating all sides, and angles, then comparing sides and angles.

$$c^2 = a^2 + b^2 - 2ab(\cos C)$$

$$(7)^2 = (8)^2 + (9)^2 - 2(8)(9)\cos C$$

$$49 = 64 + 81 - 144\cos C$$

$$49 = 145 - 144\cos C$$

$$144\cos C = 145 - 49$$

$$144\cos C = 96$$

$$\frac{144\cos C}{144} = \frac{96}{144}$$

$$\cos C = \frac{96}{144}$$

$$C = \cos^{-1}\left(\frac{96}{144}\right)$$

$$C \approx 48.2°$$

The measure of angle 'C' is the smallest, since side 'c', opposite to the angle, is the smallest side.

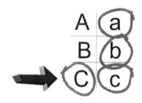

Watch the order of operations.

The '144' is attached to 'cosC' by multiplication. Division will remove the '144' correctly.

QUESTION 2

Determine the measure of the largest angle in the following triangle.

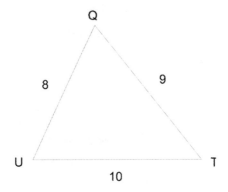

YOUR SOLUTION *Largest angle*

QUESTION 2   POSSIBLE SOLUTION   *Largest angle*

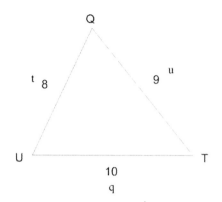

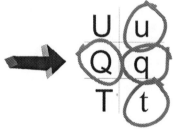

$$q^2 = u^2 + t^2 - 2ut\cos Q$$

$$(10)^2 = (9)^2 + (8)^2 - 2(9)(8)\cos Q$$

$$100 = 81 + 64 - 144\cos Q$$

$$100 = 145 - 144\cos Q$$

$$144\cos Q = 145 - 100$$

$$144\cos Q = 45$$

$$\frac{144\cos Q}{144} = \frac{45}{144}$$

$$\cos Q = \frac{45}{144}$$

$$Q = \cos^{-1}\left(\frac{45}{144}\right)$$

$$Q \approx \cos^{-1}(0.3125)$$

$$Q \approx 71.8°$$

The largest angle is angle 'Q' since it is opposite the longest side.

The triangle is not right-angled since the length of side 'u' is not equal to 6. If it was, then angle 'Q' would have measured 90° since a '6, 8, 10' triangle is a multiple of the right-angled '3, 4, 5' triangle. Since side 'u' is longer than 6 units, the measure of angle 'Q' will be less than 90°. If you want, make a model to illustrate this triangle changing from the '6, 8, 10' triangle to this one. Only one side is to change length.

Since the triangle is not right-angled, the only other choice for solving so far is the cosine law.

The measure of angle 'Q' is required', and its opposite side 'q' is available, dictating the cosine law begin with 'q'.

Watch the order of operations. Again there is a '144' is attached to 'cosQ' by a multiplication. Division will remove the '144' correctly.

QUESTION 3

Determine the measure of the largest angle in the following triangle.

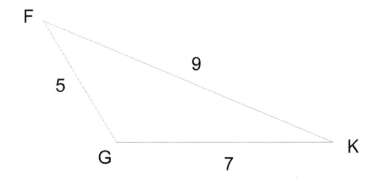

YOUR SOLUTION *Largest angle*

QUESTION 3   POSSIBLE SOLUTION   *Largest angle*

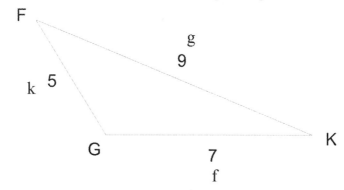

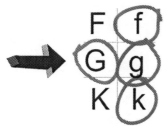

$$g^2 = f^2 + k^2 - 2fk\cos G$$

$$(9)^2 = (7)^2 + (5)^2 - 2(7)(5)\cos G$$

$$81 = 49 + 25 - 70\cos G$$

$$81 = 74 - 70\cos G$$

$$70\cos G = 74 - 81$$

$$70\cos G = -7$$

$$\frac{70\cos G}{70} = \frac{-7}{70}$$

$$\cos G = \frac{-7}{70}$$

$$G = \cos^{-1}\left(\frac{-7}{70}\right)$$

$$G = \cos^{-1}(-0.1)$$

$$G \approx 95.7°$$

The triangle is not right-angled since Pythagoras' Theorem does not work.

The area of the squares on the shorter sides.	The area of square on the longer side
$5^2 + 7^2$ $= 25 + 49$ $= 74$	$9^2 = 81$

Since the area of the square on the longest side is larger than the sum of the areas of the squares on the other 2 sides, the largest angle 'G' is obtuse, larger than 90°.

The only other choice so far is the cosine law. The measure of angle 'G' is required, and its opposite side 'g' is available, implying the cosine law begin with 'g'.

Watch the order of operations. There is a '70' is attached to 'cosG' by multiplication. Division will remove the '70' correctly.

Whenever the cosine ratio is negative, the angle will be obtuse, greater than 90°, and less than 180°.

QUESTION 4

Determine the measure of angle '*R*' in the triangle.

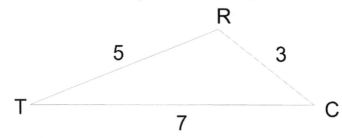

YOUR SOLUTION *Angle 'R'*

QUESTION 4   POSSIBLE SOLUTION   *Angle 'R'*

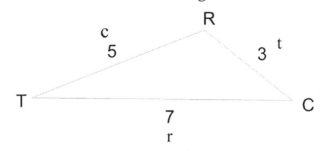

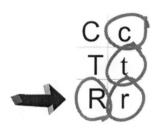

$$r^2 = c^2 + t^2 - 2ct\cos R$$

$$(7)^2 = (5)^2 + (3)^2 - 2(5)(3)\cos R$$

$$49 = 25 + 9 - 30\cos R$$

$$49 = 34 - 30\cos R$$

$$30\cos R = 34 - 49$$

$$30\cos R = -15$$

$$\frac{30\cos R}{30} = \frac{-15}{30}$$

$$\cos R = -0.5$$

$$R = \cos^{-1}(-0.5)$$

$$R = 120°$$

The triangle is not right-angled since Pythagoras' Theorem does not work.

The area of the squares on the shorter sides.	The area of square on the longer side.
$3^2 + 5^2$ $= 9 + 25$ $= 34$	$7^2 = 49$

Since the area of the square on the longest side is larger than the sum of the areas of the squares on the other 2 sides, angle 'R' is obtuse.

The only other choice so far is the cosine law. The measure of angle 'R' is required, and its opposite side 'r' is available, implying the cosine law begin with 'r'.

Watch the order of operations. There is a '30' attached to 'cosR' by multiplication. Division will remove the '30' correctly.

QUESTION 5

Determine the measure of angle '*K*' in the triangle.

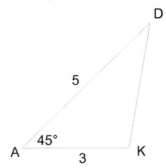

YOUR SOLUTION *Angle 'K'*

QUESTION 5   POSSIBLE SOLUTION   *Angle 'K'*

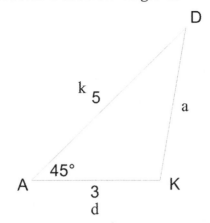

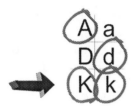

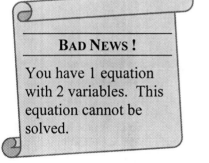

There is no guarantee of any 90° angle. The only other choice so far is the cosine law.

The measure of angle 'K' is required, and its opposite side 'k' is available, implying the cosine law begin with 'k'.

$$k^2 = a^2 + d^2 - 2ad \cos K$$

$$(5)^2 = a^2 + (3)^2 - 2a(3)\cos K$$

This time a second calculation is required.

Calculate the length of side 'a' before using this equation.

BAD NEWS !

You have 1 equation with 2 variables. This equation cannot be solved.

QUESTION 5   POSSIBLE SOLUTION   *Side 'a', then angle 'K'*

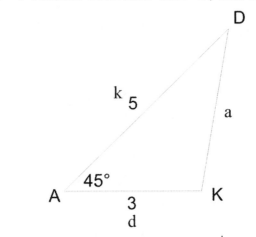

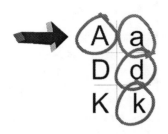

The length of side 'a' is required, and its opposite angle 'A' is available, implying the cosine law begin with 'a'.

$$a^2 = d^2 + k^2 - 2dk \cos A$$
$$a^2 = (3)^2 + (5)^2 - 2(3)(5)\cos 45°$$
$$a^2 = 9 + 25 - 30\cos 45°$$
$$a^2 \approx 34 - 21.2132$$
$$a^2 = 12.7868$$
$$a = \sqrt{12.7868}$$
$$a \approx 3.58$$

Watch the order of operations.

Store all digits from the answer, but only record these two decimal places..

QUESTION 5   POSSIBLE SOLUTION   *Angle 'K', last step*

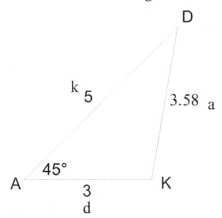

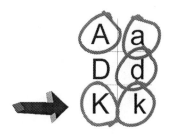

The measure of angle 'K' is required, and its opposite side 'k' is available, implying the cosine law begin with 'k'.

If possible use all digits from the previous value of 'a'.

Watch the order of operations. There is a '21.46' attached to 'cosK' by multiplication. Division will remove the '21.46' correctly.

Note: there is no alternate solution to this problem that will guarantee the correct answer faster than this method.

The cosine law is the most 'powerful' of the trig equations, but two sides are always required, and one angle.

$$k^2 = a^2 + d^2 - 2ad \cos K$$

$$(5)^2 \approx (3.58)^2 + (3)^2 - 2(3.58)(3)\cos K$$

$$25 \approx 12.79 + 9 - 21.46 \cos K$$

$$25 = 21.79 - 21.46 \cos K$$

$$21.46 \cos K = 21.79 - 25$$

$$21.46 \cos K = -3.21$$

$$\frac{21.46 \cos K}{21.46} = \frac{-3.21}{21.46}$$

$$\cos K \approx -0.1498$$

$$K = \cos(-0.1498)$$

$$K \approx 98.6°$$

QUESTION 6

Determine the length of side 'q' in the triangle.

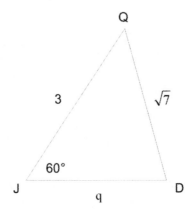

YOUR SOLUTION   *Side 'q'*

QUESTION 6   POSSIBLE SOLUTION   *Side 'q'*

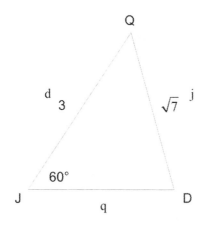

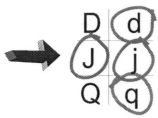

$$j^2 = d^2 + q^2 - 2dq\cos J$$

$$\left(\sqrt{7}\right)^2 = (3)^2 + q^2 - 2(3)q\cos(60°)$$

$$7 = 9 + q^2 - 6q(0.5)$$

$$7 = 9 + q^2 - 3q$$

$$0 = 9 + q^2 - 3q - 7$$

$$0 = q^2 - 3q + 2$$

$$0 = (q-2)(q-1)$$

$$q - 2 = 0$$

$$q = 2$$

or

$$q - 1 = 0$$

$$q = 1$$

The triangle is not guaranteed right-angled.

The only other choice so far is the cosine law. Angle 'J', and its opposite side 'j' are available, implying the cosine law begin with 'j'.

Since the variable 'q' has an exponent of 2, there is the possibility of two answers.

Equations of this type can be solved several ways, the most common methods: factoring, or the quadratic formula. Some calculators will also solve equations in this format.

This is not the only method to solve this question, but it is the fastest.

An accurate sketch should convince you both answers are possible.

When q=2   When q=1

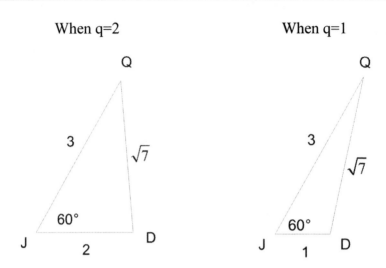

If the two sketches are placed together, the following diagram occurs.

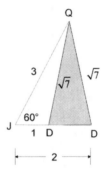

Only 2 sides were locked in position. The side with length $\sqrt{7}$ has the option of occurring angled in either direction.

Some textbooks call this the ambiguous case, but this question is simply a question with 2 possible answers.

Show Me
A Sign

Review

Determine the measure of the largest angle in the following triangle.

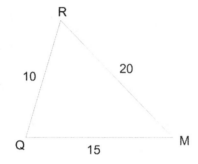

Your Solution   *Largest angle*

POSSIBLE SOLUTION *Largest angle*

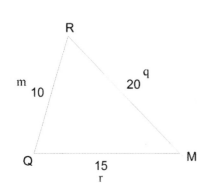

Remember to label each side with the same letter as its opposite angle, except sides use lower case. Angles use upper case.

The largest angle is opposite the longest side.

The triangle is not guaranteed right angled, so SOHCAHTOA, and Pythagoras' Theorem are not available for this triangle.

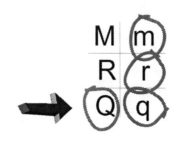

Use the cosine law with 'q' on the left side of the equation.

$$q^2 = m^2 + r^2 - 2mr\cos Q$$
$$(20)^2 = (10)^2 + (15)^2 - 2(10)(15)\cos Q$$
$$400 = 100 + 225 - 300\cos Q$$
$$400 = 325 - 300\cos Q$$
$$300\cos Q = 325 - 400$$
$$\frac{300\cos Q}{300} = \frac{-75}{300}$$
$$\cos Q = -0.25$$
$$Q = \cos^{-1}(-0.25)$$
$$Q \approx 104.5°$$

Watch the order of operations here. Since the '-300' is attached to the 'cosQ', the calculation '325 – 300' would be illegal.

REVIEW SO FAR:

1. When the triangle is right angled:
   a) Pythagoras' Theorem is available to use.
   b) the 3 basic trig ratios, sine, cosine, and tangent are available.

S	O	H
C	A	H
T	O	A

   c) $Area = \dfrac{1}{2} base \times height$

   Any other time, height was introduced *temporarily*.

2. Given two sides, and the angle *between* them:
   a) Use the *cosine law* to calculate a side length.
   b) Use $Area = \dfrac{1}{2} ab \sin C$.

3. For three sides:
   a) Use the *cosine law* to determine an angle measure.
   b) If area is required, the angle calculated in part (a) can be used
   with the formula $Area = \dfrac{1}{2} ab \sin C$.

NEW

Determine the length of side '*b*' in the following triangle.

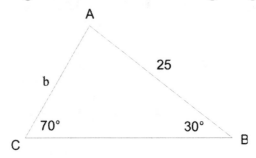

YOUR SOLUTION   *Side 'b'*

POSSIBLE SOLUTION *Side 'b'*

Since the triangle is not a 90° triangle, and the cosine law would require two sides be available, *introduce height temporarily.* This is the same starting strategy used before.

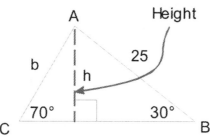

Draw the 2 new triangles separately.

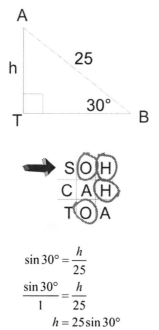

Using the height calculated from the triangle on the right side, the required side length can be calculated using the equation from the left side triangle.

Using the height, side 'b' can be calculated.

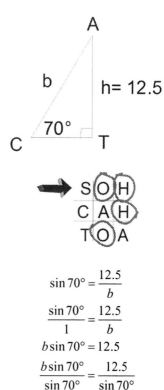

$$\sin 30° = \frac{h}{25}$$

$$\frac{\sin 30°}{1} = \frac{h}{25}$$

$$h = 25\sin 30°$$

$$h = 12.5$$

$$\sin 70° = \frac{12.5}{b}$$

$$\frac{\sin 70°}{1} = \frac{12.5}{b}$$

$$b\sin 70° = 12.5$$

$$\frac{b\sin 70°}{\sin 70°} = \frac{12.5}{\sin 70°}$$

$$b \approx 13.30$$

THEORY

In general, the previous type of problem does require some problem solving. But, you could skip this section.

Again, introduce the height temporarily.

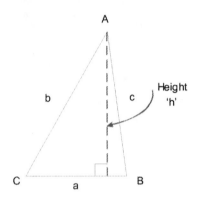

	Draw the 2 new triangles separately.	
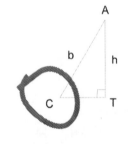	*Using the height, the lengths of side 'b' and side 'c' could be calculated.*	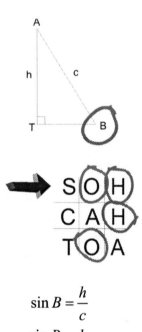
	*Both of these sine ratios can be generated, with the expectation of only original triangle letters after substitution.*	

$$\sin C = \frac{h}{b}$$

$$\frac{\sin C}{1} = \frac{h}{b}$$

$$h = b \sin C$$

$$\sin B = \frac{h}{c}$$

$$\frac{\sin B}{1} = \frac{h}{c}$$

$$h = c \sin B$$

Now, there are two equations to work with: $h = b \sin C$
and $h = c \sin B$. Since the 'h' was temporary, use substitution to
make one equation with all original triangle letters.

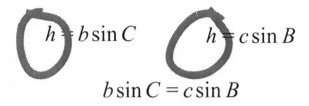

$$b \sin C = c \sin B$$

Although this format works for the equation, usually it is written in
fraction form with matching letters on each side of the equation.
With a decision to divide both sides of the equation by ' bc ', the
following occurs.

$$b \sin C = c \sin B$$

$$\frac{b \sin C}{bc} = \frac{c \sin B}{bc}$$

$$\frac{\sin C}{c} = \frac{\sin B}{b}$$

*Since these two fractions are equal, the
final equation can also be written:*

$$\frac{c}{\sin C} = \frac{b}{\sin B}$$

*The triangle can easily be relabeled to
have the formulas:*

$$\frac{a}{\sin A} = \frac{b}{\sin B} \quad or \quad \frac{c}{\sin C} = \frac{a}{\sin A}$$

These formulas are called the *sine law.*

THE SINE LAW

The fraction formed by a 'side', divided by the 'sine of its opposite angle', is equal to any other fraction, formed the same way, from the same triangle.

In a triangle with angles named U, Q, and T, and opposite sides labeled in lower case of the same letters, all of the following formulas would be true:

$$\frac{q}{\sin Q} = \frac{t}{\sin T} \qquad \frac{\sin Q}{q} = \frac{\sin T}{t}$$

$$\frac{q}{\sin Q} = \frac{u}{\sin U} \qquad \frac{\sin Q}{q} = \frac{\sin U}{u}$$

$$\frac{u}{\sin U} = \frac{t}{\sin T} \qquad \frac{\sin U}{u} = \frac{\sin T}{t}$$

Select the best formula format to use for each question.

Any new formula should be checked on a question where the answer is already available. After all, if the formula fails then, there is a huge problem. Let's redo the first 'NEW' question from this section.

QUESTION 1

Determine the length of side '*b*'.

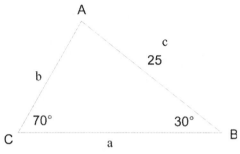

The length of side 'b' was calculated to be approximately 13.3 units in the previous solution on page 97.

YOUR SOLUTION   *Side 'b'*

QUESTION 1   POSSIBLE SOLUTION   *Side 'b'*

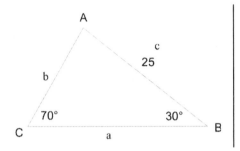

The length of side 'b' was calculated to be approximately 13.3 units on page 97.

Using a grid listing the sides, and angles, we have the following selection:

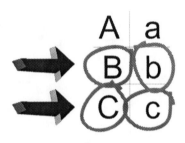

Since side 'b' is to be calculated, the best formula to use will be:

$$\frac{b}{\sin B} = \frac{c}{\sin C}$$

It is recommended the required variable be in the upper left position.

A possible solution is:

$$\frac{b}{\sin B} = \frac{c}{\sin C}$$

$$\frac{b}{\sin 30°} = \frac{25}{\sin 70°}$$

$$b\sin 70° = 25\sin 30°$$

$$\frac{b\sin 70°}{\sin 70°} = \frac{25\sin 30°}{\sin 70°}$$

$$b = \frac{25\sin 30°}{\sin 70°}$$

$$b \approx 13.30$$

Some prefer to calculate using this method instead. The answer is the same either way.

$$b = \left(\frac{25}{\sin 70°}\right)\sin 30°$$

$$b \approx 13.30$$

The sine law does work on this question, *without* calculating the height.

QUESTION 2

Determine the lengths of the remaining sides.

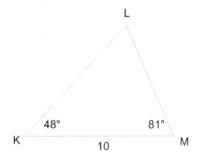

YOUR SOLUTION   *Side lengths*

QUESTION 2 POSSIBLE SOLUTION   *Side lengths*

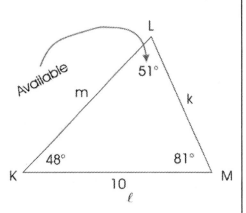

Since the sum of the measure of the angles in all triangles is 180°, the measure of angle L will be:

$$180° - 48° - 81° = 51°$$

There are two side lengths to calculate.

Select the side length to calculate first, your choice.

This solution calculates 'm' first, the longest side in this triangle.

Using a grid again, listing the sides, and angles, we have the following selection:

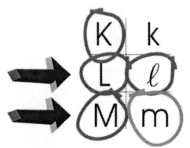

Since side 'm' is to being calculated, the best formula to use will be:

$$\frac{m}{\sin M} = \frac{l}{\sin L}$$

The required variable is recommended in the upper left corner.

A possible solution is:

$$\frac{m}{\sin M} = \frac{l}{\sin L}$$

$$\frac{m}{\sin 81°} = \frac{10}{\sin 51°}$$

$$m \sin 51° = 10 \sin 81°$$

$$\frac{m \sin 51°}{\sin 51°} = \frac{10 \sin 81°}{\sin 51°}$$

$$m = \frac{10 \sin 81°}{\sin 51°}$$

$$m \approx 12.71$$

Since 'M' is a larger angle than 'L', the side length of 'm' should be larger than '10'.

QUESTION 2   POSSIBLE SOLUTION   *Side length 'k'*

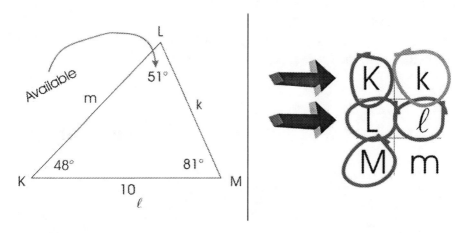

Since side 'k' is to being calculated, the best formula to use will be:

$$\frac{k}{\sin K} = \frac{l}{\sin L}$$

Try to avoid using side 'm' in any calculations unless necessary. If 'm' were used, and an error had been made calculating 'm' in the previous step, then an error would be generated in the next calculation.

A possible solution is:

$$\frac{k}{\sin K} = \frac{l}{\sin L}$$

$$\frac{k}{\sin 48°} = \frac{10}{\sin 51°}$$

$$k \sin 51° = 10 \sin 48°$$

$$\frac{k \sin 51°}{\sin 51°} = \frac{10 \sin 48°}{\sin 51°}$$

$$k = \frac{10 \sin 48°}{\sin 51°}$$

$$k \approx 9.56$$

Since 'K' is a smaller angle than 'L', the side length of 'k' should be smaller than '10'.

Since all sides and angles are available, it is a good idea to check your answers using an alternate method. Compare the relative sizes of angles, and sides by extending the original grid. This check does not guarantee your answers are perfect, just that they are realistic and recorded in the correct order.

Angle measures				Side lengths		
Small	48°	K	k	9.56	Small
Medium	51°	L	ℓ	10	Medium
Large	81°	M	m	12.71	Large

QUESTION 3

Determine the measure of angle '*E*'.

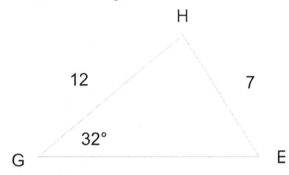

YOUR SOLUTION   *Angle 'E'*

QUESTION 3   POSSIBLE SOLUTION   *Angle 'E'*

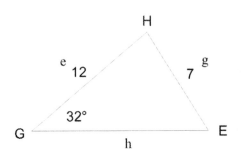

The measure of angle 'E' will be larger than 32°, since the side length of 'e' is larger than the side length of side 'g' opposite the 32°.

There is no angle holding side 'g' where it is. Maybe this side will swing to the left instead of the right. Consider this drawing option:

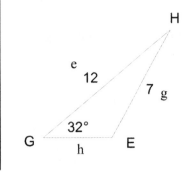

Using a grid again, we have the following selection:

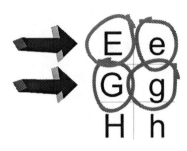

Since angle 'E' is to being calculated, the best formula to use will be:

$$\frac{\sin E}{e} = \frac{\sin G}{g}$$

The required variable is recommended in the upper left position.

A possible solution is:

$$\frac{\sin E}{e} = \frac{\sin G}{g}$$

$$\frac{\sin E}{12} = \frac{\sin 32°}{7}$$

$$7\sin E = 12\sin 32°$$

$$\frac{7\sin E}{7} = \frac{12\sin 32°}{7}$$

$$\sin E = \frac{12\sin 32°}{7}$$

$$\sin E \approx 0.9084$$

$$E \approx \sin^{-1}(0.9084)$$

$$E \approx 65.3°$$

As expected, angle 'E' is larger than 32°.

With the sine law, there is a second angle to consider. It is calculated by subtracting this answer of 65.3° from 180°.

In this case 180° - 65.3°=114.7 °.
Check entering sin 114.7° on your calculator, expecting to see the answer approximately equal to 0.9084.

Whenever the sine law is used to determine angles, there is a second angle to consider. Sometimes this second angle is rejected. This rejection will occur if the measure of the angles in the triangle will total more than 180° using the measure of the angle.

There is never a second angle to consider when calculating the measure of angles using the sine, cosine, tangent ratios (SOHCAHTOA) used with right triangles, or the cosine law.

Here are the summary diagrams for this question.

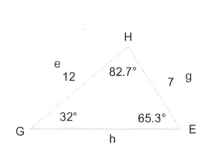
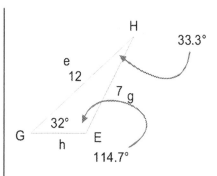

QUESTION 4

Determine the measure of angle '*Y*'.

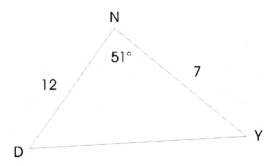

YOUR SOLUTION   *Angle 'Y'*

QUESTION 4   POSSIBLE SOLUTION   *Angle 'Y'*

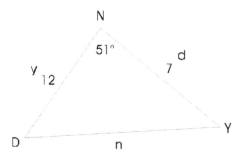

Using a grid again, we have the following selection:

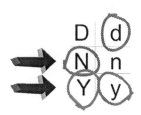

Since the measure of angle 'Y' is to be calculated, the best formula to use will be:

$$\frac{\sin Y}{y} = \frac{\sin N}{n}$$

However, the length of side 'n' is not currently available.

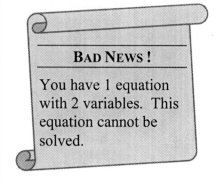

BAD NEWS !

You have 1 equation with 2 variables. This equation cannot be solved.

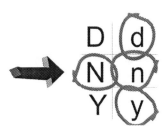

When only the given measurements are circled, two sides and the angle between them will be circled. Side 'n' can be calculated using the cosine law. This time, its length will be required in the second step of the calculation.

A possible solution is:

First, calculate the length of side 'n'.

$$n^2 = d^2 + y^2 - 2dy \cos N$$

$$n^2 = (7)^2 + (12)^2 - 2(7)(12)\cos(51°)$$

$$n^2 = 49 + 144 - 168\cos 51°$$

$$n^2 = 193 - 168\cos 51°$$

$$n^2 \approx 87.2742$$

$$n \approx \sqrt{87.2742}$$

$$n \approx 9.34$$

Since this number will be used in another calculation, keep all the decimal digits using the memory on your calculator, or the ANS (answer) key.

Now the measure of angle 'Y' can be calculated.

Using the stored value of 'n', roundoff errors will be minimized.

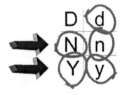

$$\frac{\sin Y}{y} = \frac{\sin N}{n}$$

$$\frac{\sin Y}{12} = \frac{\sin 51°}{9.34}$$

$$\sin Y = \frac{12\sin 51°}{9.34}$$

$$\sin Y \approx 0.9983$$

$$Y \approx \sin^{-1}(0.9983)$$

$$Y \approx 86.6°$$

or

$$Y \approx 180° - 86.6°$$

$$Y \approx 93.4°$$

As usual, when calculating an angle using the sine law, there are two answers to consider. In this example there are two possible triangles to consider.

But, the sides were locked in position since the angle was between the two given sides. Only one triangle is possible. If you made a construction of this triangle, only this diagram is possible.

*The sine law will **not** decide the correct answer; neither will the small, medium, large test. The only guarantee of the correct angle will be from the cosine law.*

The final check using all decimals:

$$y^2 = d^2 + n^2 - 2dn \cos Y$$
$$12^2 \approx 7^2 + 9.34^2 - 2(7)(9.34)\cos Y$$
$$144 \approx 49 + 87.2742 - 130.7889 \cos Y$$
$$144 = 136.2742 - 130.7889 \cos Y$$
$$130.7889 \cos Y = 136.2742 - 144$$
$$130.7889 \cos Y = -7.7258$$
$$\frac{130.7889 \cos Y}{130.7889} = \frac{-7.7258}{130.7889}$$
$$\cos Y \approx -.0591$$
$$Y \approx \cos^{-1}(-.0591)$$
$$Y \approx 93.4°$$

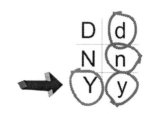

Since 'cosY' is negative, the measure of angle 'Y' will be obtuse. This means 93.4° will be the correct answer for the measure of the angle without completing this calculation.

Be careful with the sine law when angles are involved. There will be 2 angles to consider. The cosine law is a superior method, but requires 2 sides and *any* angle. *Ideally*, the required side is opposite an available angle. But as we saw in an earlier session, this is not a requirement.

SUMMARY
SHEETS

The shaded area lists the method(s) used to 'solve a triangle'. When a triangle is 'solved', the measures of all sides and all angles will be known.

Number Of Sides \ Number Of Angles	0	1	2	3
0	You are not ready yet. Go get some information.	At least one side is required. Go get more information.	One side still needed, but the 3rd angle can be calculated.	At least one side is required.
1	You need some more information.	Get another angle or side.	**Sine Law** Calculate the 3rd angle.	**Sine Law**
2	Another angle or another side is needed.	If there is no side length on a side opposite the angle **Cosine Law** If there is an angle opposite one of the sides **Sine Law** (Cosine Law is safer.)	**Sine Law** Calculate the 3rd angle. **Cosine Law** (Optional)	**Sine Law** Cosine Law (Optional)
3	**Cosine Law**	**Sine Law** (Cosine Law is safer.)	Calculate the 3rd angle.	You are done!

Notes:

1. The longest side is always opposite the largest angle, and the shortest side is always opposite the smallest angle, etc. Make a choice to check your final answers for this simple S, M, L test. Of course this test will not guarantee your answer is perfect, just that it is reasonable.

2. Any information for a triangle with a 90° angle is expected to be solved using "SOHCAHTOA" (the method from the beginning of your work with ratios). At least one of the cosine law and the sine law will always work in any type of triangle.

3. The *sine law* is excellent when a side is being calculated (and 2 angles are given), but when used to calculate an angle, most calculators have decided to only give one of the 2 possible answers to consider. The calculator may even give the wrong answer for that application. Remember for the sine law, the other angle to consider is calculated by subtracting the calculator answer for the angle from 180°. Then a decision is required to determine the correct answer(s). This is never a problem for the cosine law, so it can be used to test the 2 possible answers.

4. The *cosine law* is the most powerful of the formulas, but sometimes power is not necessary. For example, using a sledgehammer to put a nail in the wall for a picture, is not a good application for the sledgehammer, but it will work. (The sledgehammer may even cause some other problems.)

5. A triangle is considered *solved* when all 3 sides, and all 3 angles, are available, but usually it is not necessary to complete all the calculations.

6. The minimum information required for any triangle is one side, and any other 2 measurements. Often though, any application outside of school problems has extra information. This extra information can be used to check the accuracy of supplied information.

7. The basic trig ratios are sine, cosine, and tangent (the ones on the calculator). There are three others that are the reciprocals of these, but as you can see so far, they were not necessary. If they were really important, there would be additional buttons on the calculator.

Now for the big moment. You have access to all the skills required to solve any problem where a triangle is involved. Even triangles that are three dimensional will follow the same basic rules. You may want to refer to the summary chart when working on these problems, but with practise, you will soon not need to refer to the chart.

The next sessions contain a series of standard questions. The solution to each question, with comments, will be on one side of the book, with room for your work on the other side. Fold the book so you can't see the solution, or at least … cover the solution.

MULTI-STEP QUESTIONS

QUESTION 1

Determine the area of triangle ABC.

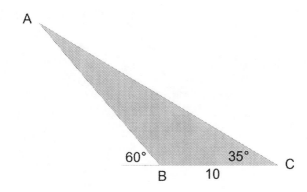

YOUR SOLUTION   *Area*

QUESTION 1   POSSIBLE SOLUTION   *Area*

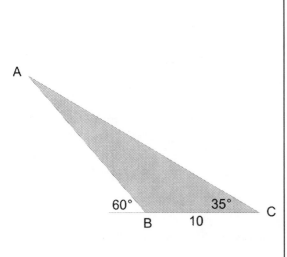

Solution 1

Since only one side length is given, consider calculating the height, and then use the area formula.

$$Area = \frac{1}{2}base \times height$$

Solution 2

Determine the length of side AC, and use

$$Area = \frac{1}{2}ab \sin C$$

developed earlier.

Since the height solution requires the steps of solution 2, it will be completed first. It is a faster solution.

SOLUTION 2

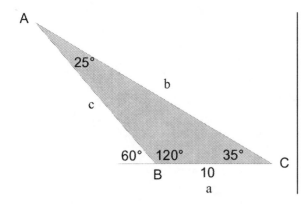

All angles are available inside the original triangle.

Area Solution:
1. *Determine the length of side 'b'.*
2. *Calculate the area.*

Determine the length of side 'b'.

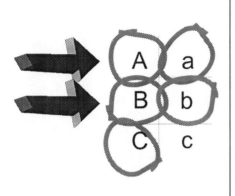

Using the *sine law*:

$$\frac{b}{\sin B} = \frac{a}{\sin A}$$

$$\frac{b}{\sin 120°} = \frac{10}{\sin 25°}$$

$$b\sin 25° = 10\sin 120°$$

$$\frac{b\sin 25°}{\sin 25°} = \frac{10\sin 120°}{\sin 25°}$$

$$b = \frac{10\sin 120°}{\sin 25°}$$

$$b \approx 20.4919$$

Now, calculate the area.

$$Area = \frac{1}{2}ab\sin C$$

$$Area = \frac{1}{2}(10)(20.4919)\sin 35°$$

$$Area \approx 58.77 \text{ units}^2$$

Keep all decimals until the last step to maximize accuracy. Use the calculator's memory, or write out the decimals.

The length of side 'c' is not required, but it can easily be calculated using the *sine law* again.

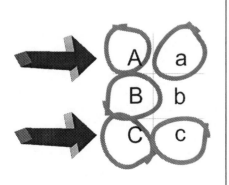

$$\frac{c}{\sin C} = \frac{a}{\sin A}$$

$$\frac{c}{\sin 35°} = \frac{10}{\sin 25°}$$

$$c\sin 25° = 10\sin 35°$$

$$\frac{c\sin 25°}{\sin 25°} = \frac{10\sin 35°}{\sin 25°}$$

$$c = \frac{10\sin 35°}{\sin 25°}$$

$$c \approx 13.57$$

SOLUTION 1   *Calculate height, then the area.*

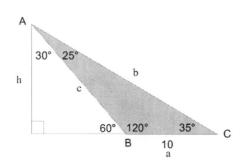

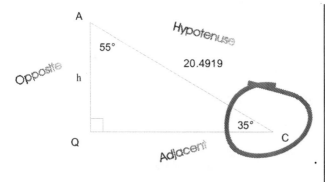

Side '*b*', the hypotenuse, is available from the previous solution.

$$b=20.4919$$

Using the largest triangle of the three, the following occurs.

$$\sin 35° = \frac{h}{20.4919}$$
$$h = 20.4919 \sin 35°$$
$$h \approx 11.7537$$

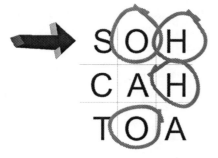

The final step, complete the area calculation.

$$Area = \frac{1}{2} base \times height$$

$$Area = \frac{1}{2}(10)(11.7537)$$

$$Area \approx 58.77 \text{ units}^2$$

Keep all decimals until the last step to maximize accuracy. Use the calculator's memory, or write out the decimals.

QUESTION 2

Determine the height of the following post. (The diagram is 3 dimensional.)

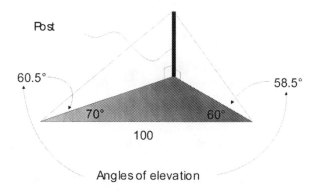

YOUR SOLUTION   *Height*

QUESTION 2   POSSIBLE SOLUTION   *Height*

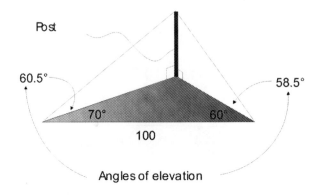

Post

60.5°

58.5°

70°

60°

100

Angles of elevation

There are 4 triangles that could be worked with, but not all are necessary. A possible method:

1. Calculate one side from the bottom (shaded) triangle.
2. Use a second triangle containing the side from the first step, and calculate the post height.

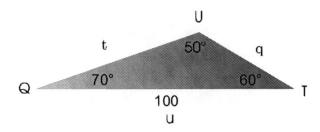

U

t

50°

q

70°

60°

Q

100

T

U

Either the length of side 'q', or side 't' can be calculated. The data provided in the original diagram can be checked by calculating both lengths. In fact, if the post height were critical, the measurement of both angles of elevation permits a check for accuracy of all measurements.

Solving for 't':

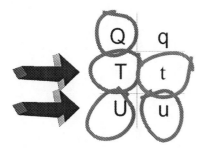

$$\frac{t}{\sin T} = \frac{u}{\sin U}$$

$$\frac{t}{\sin 60°} = \frac{100}{\sin 50°}$$

$$t = \frac{100 \sin 60°}{\sin 50°}$$

$$t \approx 113.05$$

Solving for 'q':

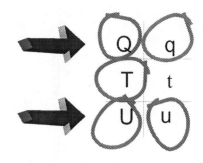

$$\frac{q}{\sin Q} = \frac{u}{\sin U}$$

$$\frac{q}{\sin 70°} = \frac{100}{\sin 50°}$$

$$q = \frac{100 \sin 70°}{\sin 50°}$$

$$q \approx 122.67$$

These answers pass the small, medium, large test.

large	70°	Q	q	122.67	large
medium	60°	T	t	113.05	medium
small	50°	U	u	100	small

Both answers will be used in another calculation. Complete accuracy is preferred. Ideally, store the answers in your calculator's memory. If necessary record all digits on paper, then continue.

Using 't', and the left side triangle:

Using 'q', and the right side triangle:

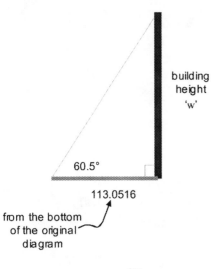

building
height
'w'

60.5°

113.0516

from the bottom
of the original
diagram

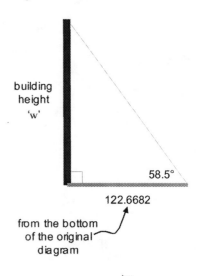

building
height
'w'

58.5°

122.6682

from the bottom
of the original
diagram

$$\tan 60.5° = \frac{w}{113.0516}$$

$$\frac{\tan 60.5°}{1} = \frac{w}{113.0516}$$

$$w = 113.0516 \tan 60.5°$$

$$w \approx 199.82$$

$$\tan 58.5° = \frac{w}{122.6682}$$

$$\frac{\tan 58.5°}{1} = \frac{w}{122.6682}$$

$$w = 122.6682 \tan 58.5°$$

$$w \approx 200.18$$

Using both methods, the separate answers are very close to 200 units. If this height were critical, the original diagram supplies sufficient information to allow a check on the information collected, using a second set of data.

APPLICATIONS

The roof structures for newer homes are prefabricated. They are then installed on site, using a crane. A calculation is required to determine the height of this roof, from the peak, or "ridge", to the top of the second floor.

This photo shows the results after construction has been completed.

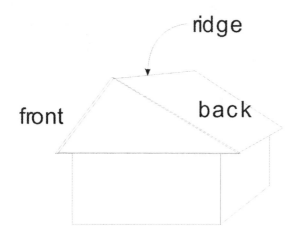

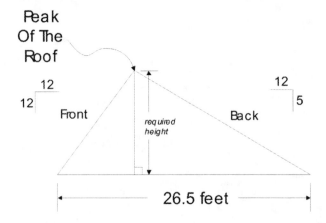

The numbers on the front and back roof sections come from the blueprints for the house and are used to indicate the pitch of the roof. They are just slope indicators, stating the ratios of vertical and horizontal distances. From the first topic, slope can be used to calculate an angle using the *tangent* ratio.

The ridge of a roof is the horizontal beam across the top of the roof. This peak height needs to be accurate to guarantee the slope of each roof section as shown.

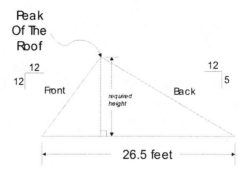

YOUR SOLUTION   *Height of the peak*

POSSIBLE SOLUTION   *Height of the peak*

1. Calculate the angle of the roof at the front, and back of the house.
2. The truss company, or the contractor, will require the lengths for both sections of the roof, so calculate both roof lengths now.
3. Calculate the height of the peak.
4. Check your work.

Angle Calculations:

FRONT | **BACK**

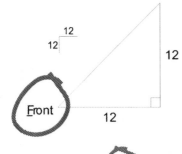

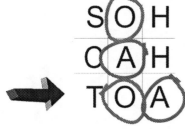

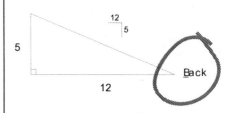

$$\tan F = \frac{12}{12}$$

$$F = \tan^{-1}\left(\frac{12}{12}\right)$$

$$F = 45°$$

$$\tan B = \frac{5}{12}$$

$$B = \tan^{-1}\left(\frac{5}{12}\right)$$

$$B = 22.62°$$

No calculations were necessary since the right-angled triangle was isosceles.

Keep this number exact using the memory on your calculator.

Roof Lengths

The angle at the peak is available by calculating $180° - 45° - 22.62°$ since the 3 angle measures must total $180°$.

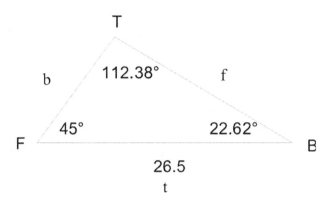

FRONT:

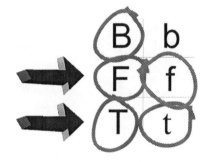

$$\frac{f}{\sin F} = \frac{t}{\sin T}$$

$$\frac{f}{\sin 45°} = \frac{26.5}{\sin 112.38°}$$

$$f = \frac{26.5 \sin 45°}{\sin 112.38°}$$

$$f \approx 20.26$$

BACK:

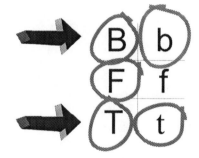

$$\frac{b}{\sin B} = \frac{t}{\sin T}$$

$$\frac{b}{\sin 22.62°} = \frac{26.5}{\sin 112.38°}$$

$$b = \frac{26.5 \sin 22.62°}{\sin 112.38°}$$

$$b \approx 11.02$$

Both roof sides have been calculated giving you a choice of calculating the height using the front, or back triangle.

Since the roof calculations are important, do them both.

Since both triangles are right-angled, using the basic trig ratios is the faster method. In both cases the hypotenuse is available, and the opposite side is required when the angles used are the angles of elevation calculated earlier.

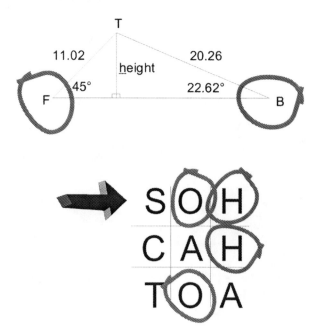

$$\sin 45° = \frac{height}{11.02}$$
$$height = 11.02\sin 45°$$
$$height \approx 7.79$$

$$\sin 22.62° = \frac{height}{20.26}$$
$$height = 20.26\sin 22.62°$$
$$height \approx 7.79$$

Since the units were given in feet, these lengths will also be in feet.

Although 26.5 feet is available on a tape measure, 26 feet, 6 inches, the height of 7.79 is not.

There are 12 inches in a foot, with inches subdivided to a 16th, or a 32nd, for a carpenter.

To change 7.79 to a useful number:
1. Multiply the decimal portion by 12 to calculate the number of inches.
2. Multiply the next decimal portion by 32 to calculate the number of 32nds.

Original Length
7.79

Number of Feet
7

Number of Inches Remaining
$0.79 \times 12 = 9.48$

Number of 32nds Remaining

$0.48 \times 32 = 15$
Answer

$$7.79 = 7'9\frac{15}{32}"$$

Note:
If all the decimals are kept at every step,
there is an $\frac{1}{8}$th of an inch difference.

$$7.79 = 7'9\frac{17}{32}"$$

ANGLE FORMATS

The examples in this book have used angles in decimal format. Surveyors, and others, will begin with angles in a different format. The change between methods is comparable to changes between metric and Imperial.

In both cases, some calculators will make these changes with the press of a button. If necessary, the angles can be adjusted using the following method.

On most calculators, the required button will fit one of the following 3 formats:
DMS,
DDMMSS,
or °'"

Degree measurements can be subdivided into minutes, and seconds, rather than decimals. Each degree has 60 minutes, the same as an hour has 60 minutes. Each minute has 60 seconds. No surprise there. One degree then, has $60 \times 60 = 3600$ seconds.

EXAMPLE 1 *Changing to decimal degrees*

An example of exact equivalent angles would be $42°30' = 42.5°$. An angle measured as $61°23'57''$ should be approximately 61.5°, but a bit less.

Use the calculation:

$$61°23'57'' = 61 + \frac{23}{60} + \frac{57}{3600}$$
$$= 61.39916667$$
$$\approx 61.40°$$

$$23' = \left(\frac{23}{60}\right)°$$

$$57'' = \left(\frac{57}{3600}\right)°$$

EXAMPLE 2   *Changing to degrees, minutes, and seconds*

In reverse, 61.4° should be very close to 61°23′57″. To determine the minutes, and seconds, only the decimal parts are used.

Calculating minutes, use $0.4° \times 60 = 24′$, which is very close to 61°23′57″.

To generate the best calculator answer, work with the full decimal angle 61.39916667°.

Again, only use the decimal part:

For minutes	For the seconds (only use the decimal part):
$0.39916667° \times 60 = 23.9500002′$	$0.9500002′ \times 60 \approx 57.00″$

This gives us $61.39916667° \approx 61°23′57″$.

The answer is read: 61 degrees, 23 minutes, and 57 seconds.

QUESTION 1

Determine the angle formed by a slope of $\dfrac{4}{3}$ in the format: "degrees, minutes, and seconds (DEG MIN SEC)."

YOUR SOLUTION

QUESTION 1   POSSIBLE SOLUTION

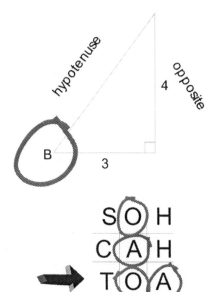

$$\tan B = \frac{4}{3}$$

$$B = \tan^{-1}\left(\frac{4}{3}\right)$$

$$B \approx 53.13010235°$$

This question was solved with a de... answer for the measure of the angl... earlier chapters.

The slope is the tangent ratio.

Minute Calculation:
Only use the decimal part of the last number.
$$0.13010235 \times 60' = 7.806141249'$$

Second Calculation:
Only use the decimal part of the last number.
$$0.806141249 \times 60'' = 48.37''$$

Decimal Answer:
53.13°

Other Format:
$$53°7'48.37'' \; \left(53°7'48''\right)$$

QUESTION 2
Write the angle $22°19'53''$ in decimal form.

YOUR SOLUTION

QUESTION 2   POSSIBLE SOLUTION

$$22°19'53'' = 22 + \frac{19}{60} + \frac{53}{3600}$$
$$= 22.33138889°$$
$$\approx 22.3°$$

SAMPLES

QUESTION 1

Determine the length of side '*q*'.

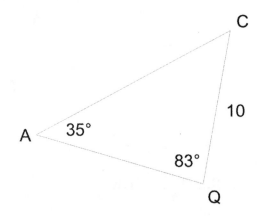

YOUR SOLUTION   *Side 'q'*

QUESTION 1   POSSIBLE SOLUTION   *Side 'q'*

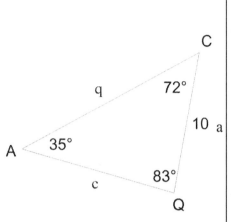

Thinking "stuff":

✓ The triangle has 2 angles, and 1 side given, the minimum amount of information required.

✓ The 3rd angle measures 72° (triangle angles total 180°), so, the triangle is not right angled. The formula to use is one of *sine law*, or *cosine law*.

✓ The angle 'A', and its opposite side 'a' are available. Use the *sine law*.

✓ The length of side 'q' is guaranteed longer than 10 units since angle 'Q' is the largest angle.

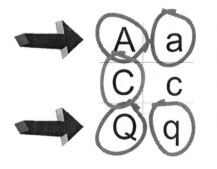

$$\frac{q}{\sin Q} = \frac{a}{\sin A}$$

$$\frac{q}{\sin 83°} = \frac{10}{\sin 35°}$$

$$q = \frac{10\sin 83°}{\sin 35°} \text{ (optional step)}$$

$$q \approx 17.3$$

SUMMARY

S	A=35°	a=10	S
M	C=72°		
L	Q=83°	q≈17.3	L

The basic <u>s</u>mall, <u>m</u>edium, <u>l</u>arge test suggests the 17.3 unit length is reasonable.

QUESTION 2

Determine the length of side '*k*'.

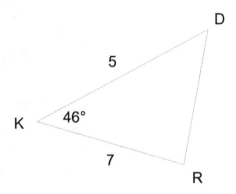

YOUR SOLUTION   *Side 'k'*

QUESTION 2   POSSIBLE SOLUTION   *Side 'k'*

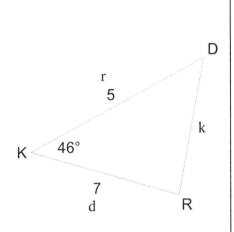

Thinking "stuff":

✓ Probably the *cosine law* is required since that last formula was the *sine law.* ☺
✓ The triangle has 2 sides, and 1 angle given, the minimum amount of information required.
✓ The triangle is not given as right angled, so, the formula to use is one of *sine law*, or *cosine law.*
✓ No given side is opposite the 46° angle. Use the *cosine law.*

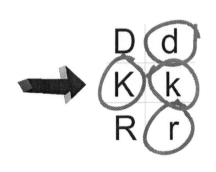

$$k^2 = d^2 + r^2 - 2dr\cos K$$
$$k^2 = (7)^2 + (5)^2 - 2(5)(7)\cos 46°$$
$$k^2 = 49 + 25 - 70\cos 46°$$
$$k^2 = 74 - 70\cos 46°$$
$$k^2 = 74 - 70(0.6947)$$

Watch the order of operations

$$k^2 = 25.3739$$
$$k \approx 5.04$$

Some of these steps do not need to be written down, but at least include the original formula with the numbers substituted.

The SML test (small, medium, large test) is not useful this time since only 1 angle is available.

QUESTION 3

Determine the measure of angle '*B*'.

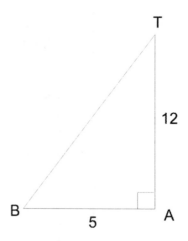

YOUR SOLUTION   *Angle 'B'*

Question 3 Possible Solution *Angle 'B'*

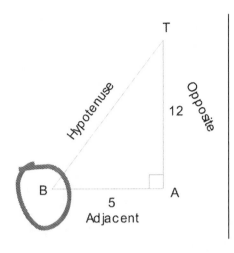

Thinking "stuff":

✓ This time there is a 90° angle.
✓ The triangle has 2 sides, and 1 angle given, the minimum required info.
✓ The only '1 step' solution is to use the *tangent* ratio (slope).

These ratios are available since the triangle has a 90° angle.

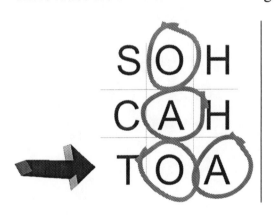

$$\tan B = \frac{12}{5}$$

$$B = \tan^{-1}\left(\frac{12}{5}\right)$$

$$B \approx 67.4°$$

QUESTION 4

Determine the measure of the largest angle.

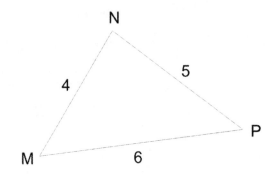

YOUR SOLUTION   *Largest angle*

QUESTION 4   POSSIBLE SOLUTION   *Largest angle*

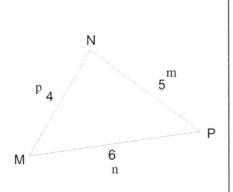

Thinking "stuff":

✓ The largest angle is opposite the longest side, so, angle 'N' is the largest angle.

✓ The triangle has 3 sides, again, the minimum info required.

✓ Although a 3, 4, 5 triangle has a 90° angle, this triangle with each side 1 unit longer has a different size for the largest angle.

✓ The *cosine law* is required.

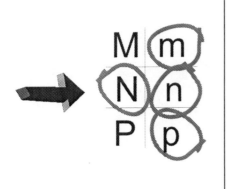

$$n^2 = m^2 + p^2 - 2mp \cos N$$
$$6^2 = 5^2 + 4^2 - 2(5)(4) \cos N$$
$$36 = 25 + 16 - 40 \cos N$$
$$36 = 41 - 40 \cos N$$
$$40 \cos N = 41 - 36$$
$$40 \cos N = 5$$
$$\frac{40 \cos N}{40} = \frac{5}{40}$$
$$\cos N = 0.125$$
$$N = \cos^{-1}(0.125)$$
$$N \approx 82.8°$$

The measure of the largest angle will be 82.8°. Since only one angle I available, the SML test is not a worthwhile test. (The smallest measure possible for the largest angle is 60°, and this 82.8° angle is larger.)

QUESTION 5

Determine the length of side '*w*'.

YOUR SOLUTION   *Side 'w'*

QUESTION 5   POSSIBLE SOLUTION   *Side 'w'*

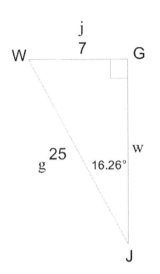

Thinking stuff:

✓ This time there is extra information giving more than one possible method.
✓ Method 1 → Pythagoras' Theorem (since there is a 90° angle).
✓ Method 2 → Select the 16.26° angle, the hypotenuse, and CAH ratio.
✓ Method 3 → determine the 3rd angle, and use the *sine law*.

METHOD 1	METHOD 2	METHOD 3

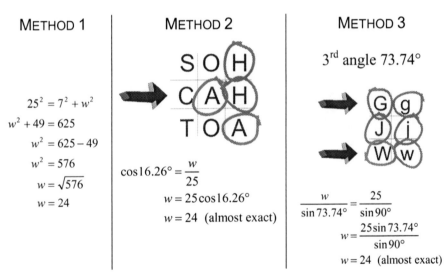

METHOD 1

$$25^2 = 7^2 + w^2$$
$$w^2 + 49 = 625$$
$$w^2 = 625 - 49$$
$$w^2 = 576$$
$$w = \sqrt{576}$$
$$w = 24$$

METHOD 2

S O H
C A H
T O A

$$\cos 16.26° = \frac{w}{25}$$
$$w = 25\cos 16.26°$$
$$w = 24 \quad \text{(almost exact)}$$

METHOD 3

3rd angle 73.74°

G g
J j
W w

$$\frac{w}{\sin 73.74°} = \frac{25}{\sin 90°}$$
$$w = \frac{25\sin 73.74°}{\sin 90°}$$
$$w = 24 \quad \text{(almost exact)}$$

The length of 24 units is less than the hypotenuse, and opposite the middle sized angle, making 24 units a reasonable length.

QUESTION 6

Determine the measure of angle '*V*'.

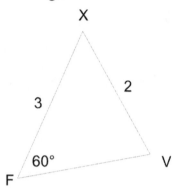

YOUR SOLUTION   *Angle 'V'*

QUESTION 6   POSSIBLE SOLUTION   *Angle 'V'*

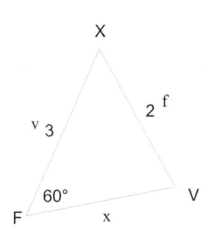

Thinking …..

✓ Minimum information given: 3 measurements, including at least one side.
✓ No 90° angle guaranteed … only *sine law* or *cosine law* available.
✓ A side and opposite angle are available → use the *sine law*.
✓ Angle 'V' is expected to be larger than 60° …

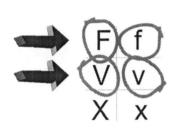

$$\frac{\sin V}{v} = \frac{\sin F}{f}$$

$$\frac{\sin V}{3} = \frac{\sin 60°}{2}$$

$$\sin V = \frac{3\sin 60°}{2}$$

$$\sin V = 1.299$$

$$V = \sin^{-1}(1.299)$$

NOT POSSIBLE!!!

The calculator will not give an answer since the largest possible value for the sine ratio, and the cosine ratio, is 1.0. (The hypotenuse length is in the denominator of the ratio and is always a longer side than the numerator, by definition.) This means it is not possible to construct this triangle using a scale drawing. The side attempting to join X, and V is not long enough. Try making a scale drawing.

QUESTION 7

Although this diagram is still only a sketch, the side with length of 2 units will not reach to the other side of the triangle. It's not even locked in position by an angle, but can swing as shown. Determine the new length of the side so it will reach side FV and make an angle of 90° with that side. (This new line will be an altitude from point X. Its length will be the height of triangle XFQ.)

ORIGINAL (FROM QUESTION 6)　　　REQUIRED LENGTH XQ

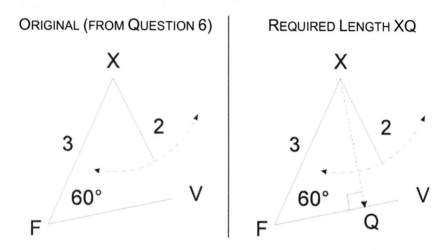

YOUR SOLUTION　　*Length of 'XQ'*

QUESTION 7   POSSIBLE SOLUTION   *Length of 'XQ'*

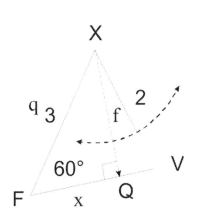

Thinking again …

✓ The new length will be less than 3 and more than 2 units.

✓ This triangle will have 1 side, and all angles available.

✓ Method 1: the triangle has a 90° angle so SOH is the ratio to use.

✓ Method 2: the *sine law* can be used.

METHOD 1

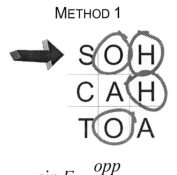

$$\sin F = \frac{opp}{hyp}$$

$$\sin 60° = \frac{f}{3}$$

$$f = 3\sin 60°$$

$$f \approx 2.60$$

METHOD 2

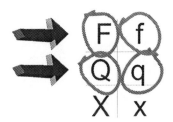

$$\frac{f}{\sin F} = \frac{q}{\sin Q}$$

$$\frac{f}{\sin 60°} = \frac{3}{\sin 90°}$$

$$f = \frac{3\sin 60°}{\sin 90°}$$

$$f \approx 2.60$$

This length falls in the expected range. Any length less, and the triangle will never be possible. Anything longer and there could be two triangles. (Refer back to the *sine law* where angles were calculated.)

QUESTION 8

Determine the height of the triangles, *side TB*.

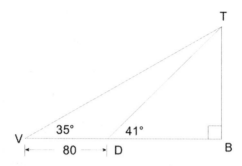

YOUR SOLUTION   *Height*

QUESTION 8   POSSIBLE SOLUTION   *Height*

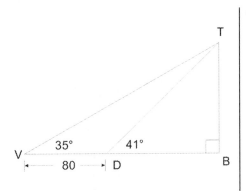

Thinking …

✓ There are 3 visible triangles, and the height is in 2 of them.
✓ The side TD is in 2 of the triangles.
✓ The angle beside the 41° angle is 139°. (Subtract 41° from 180°.)

Usually I prefer to see the 3 triangles separately to get a full picture.

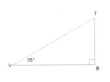
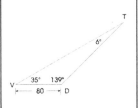
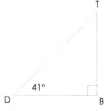

✓ Only 2 angles available.
✓ Side VT could be calculated using the next triangle.
✓ Height can be calculated using the SOH ratio.

✓ A side, and 3 angles available → *sine law*
✓ TD is a shared side → calculate it for use on the next triangle.

✓ Only 2 angles available.
✓ Side TD is shared, and available from the previous triangle.
✓ Height can be calculated using the SOH ratio

USING THE MIDDLE DIAGRAM, THEN THE THIRD:

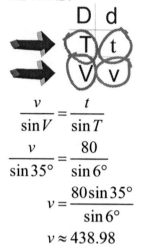

$$\frac{v}{\sin V} = \frac{t}{\sin T}$$

$$\frac{v}{\sin 35°} = \frac{80}{\sin 6°}$$

$$v = \frac{80\sin 35°}{\sin 6°}$$

$$v \approx 438.98$$

Now, use the small triangle (the 3rd one), with this side.

$$\sin D = \frac{opp}{hyp}$$

$$\sin 41° = \frac{opp}{438.98}$$

$$opp = 438.98\sin 41°$$

$$opp \approx 288.0$$

All decimals were kept using the calculator's memory. Any rounding that occurs in the solution reduces the accuracy of the final answer.

USING THE MIDDLE DIAGRAM, THEN THE FIRST.

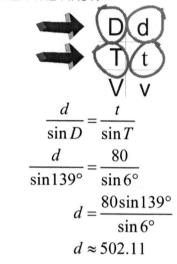

$$\frac{d}{\sin D} = \frac{t}{\sin T}$$

$$\frac{d}{\sin 139°} = \frac{80}{\sin 6°}$$

$$d = \frac{80\sin 139°}{\sin 6°}$$

$$d \approx 502.11$$

Now, use the large triangle (the 1st one), with this side.

$$\sin V = \frac{opp}{hyp}$$

$$\sin 35° = \frac{opp}{502..11}$$

$$opp = 502.11\sin 35°$$

$$opp \approx 288.0$$

All decimals were kept using the calculator's memory. Any rounding that occurs in the solution reduces the accuracy of the final answer.

WHOSE EQUATION IS IT ANYWAYS ? PART 2

CALCULATOR REVIEW

Calculate the answer(s) to the following equations. Side lengths are to be accurate to 2 decimal places, and angle measurements are to be accurate to 1 decimal place.

$\sin 80° = \dfrac{y}{5}$	
$\cos 37° = \dfrac{10}{r}$	
$\tan B = 4$	
$\dfrac{c}{\sin 25°} = \dfrac{32}{\sin 32°}$	
$d^2 = 5^2 + 7^2 - 2(5)(7)\cos 48°$	
$\dfrac{\sin E}{10} = \dfrac{\sin 20°}{30}$	
$20^2 = 12^2 + 16^2 - 2(12)(16)\cos Q$	

CALCULATOR SOLUTIONS

$\sin 80° = \dfrac{y}{5}$	$\sin 80° = \dfrac{y}{5}$ $\dfrac{\sin 80°}{1} = \dfrac{y}{5}$ $y = 5\sin 80°$ $y \approx 4.92$ *Since the hypotenuse has a length of 5, this answer will be less than that length. The 'y' is from an opposite side, and its length must be less than the hypotenuse.*
$\cos 37° = \dfrac{10}{r}$	$\cos 37° = \dfrac{10}{r}$ $\dfrac{\cos 37°}{1} = \dfrac{10}{r}$ $r\cos 37° = 10$ $\dfrac{r\cos 37°}{\cos 37°} = \dfrac{10}{\cos 37°}$ $r \approx 12.52$ *The length is expected longer than 10 since 'r' is the length of the hypotenuse. It is expected to be the longest side.*
$\tan B = 4$	$\tan B = 4$ $B = \tan^{-1}\left(4\right)$ $B \approx 76.0°$ *Although each of the sine and cosine ratios cannot be larger than 1, the tangent ratio does not have this restriction. This is the correct angle measure for the placement of a ladder. (Refer to page 8, example 2.)*
$\dfrac{c}{\sin 25°} = \dfrac{32}{\sin 32°}$	$\dfrac{c}{\sin 25°} = \dfrac{32}{\sin 32°}$ $c = \dfrac{32\sin 25°}{\sin 32°}$ $c \approx 25.52$ *Since side 'c' is opposite a smaller angle than the '32°', its length is expected to be shorter than the side length of '32'. Since sine is also a ratio, 'c' will not equal exactly 25.*

$$d^2 = 5^2 + 7^2 - 2(5)(7)\cos 48°$$

$$d^2 = 5^2 + 7^2 - 2(5)(7)\cos 48°$$
$$d^2 = 25 + 49 - 70\cos 48°$$
$$d^2 = 74 - 70\cos 48°$$
$$d^2 \approx 74 - 70(0.6691)$$
$$d^2 \approx 74 - 46.8361$$
$$d^2 = 27.1609$$
$$d = \sqrt{27.1609}$$
$$d \approx 5.21$$

Using the extreme cases of a flat triangle, the 3rd side length cannot be longer than 12 (when the angle measure between the sides is 180°), and the side length cannot be less than 2 units (when the angle measure between them is 0°). With the side lengths of 5, and 7, the third side must be between 2 and 12, otherwise the triangle cannot be drawn.

$$\frac{\sin E}{10} = \frac{\sin 20°}{30}$$

$$\frac{\sin E}{10} = \frac{\sin 20°}{30}$$
$$\sin E = \frac{10\sin 20°}{30}$$
$$\sin E \approx 0.1140$$
$$E = \sin^{-1}(0.1140)$$
$$E \approx 6.5°$$

There are 2 possible angle measures from the sine law. This answer is the calculator answer. The other one is 180° – 6.5° = 173.5°.

However, the question has an angle of 20° from the same triangle. The 173.5° angle would not be possible since the measures of the triangle angles would then total more than 180°

$20^2 = 12^2 + 16^2 - 2(12)(16)\cos Q$	$20^2 = 12^2 + 16^2 - 2(12)(16)\cos Q$ $400 = 144 + 256 - 384\cos Q$ $400 = 400 - 384\cos Q$ $400 - 384\cos Q = 400$ $-384\cos Q = 0$ $\dfrac{-384\cos Q}{-384} = \dfrac{0}{-384}$ $\cos Q = 0$ $Q = \cos^{-1}(0)$ $Q = 90°$ *This answer could have been recognized earlier, even before calculating. The 12, 16, 20 triangle is 4 times the size of the famous right-angled 3, 4, 5 triangle discussed earlier. (Refer to page 5.)*

EQUATION REVIEW

State the best equation for the following situations.

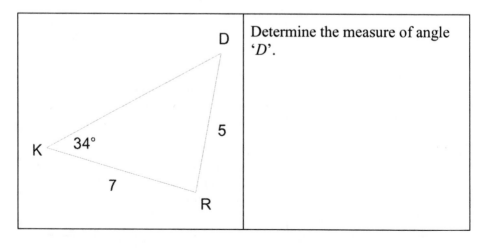

	Determine the measure of angle '*D*'.

X 3 60° T Z 2	Determine the length of side '*z*'.
T 57° B A 5	Determine the length of side '*a*'.
T 3 2 Q 4 W	Determine the measure of the smallest angle.
Q 5 B 10 A	Determine the length of side '*a*'.

EQUATION SOLUTIONS

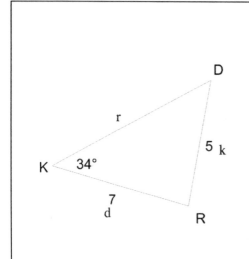

Angle 'D'

This triangle is not a 90° triangle, limiting options to sine law, or cosine law.

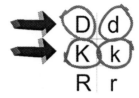

$$\frac{\sin D}{7} = \frac{\sin 34°}{5}$$

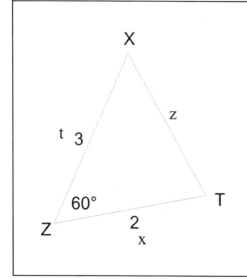

Side 'z'

This triangle is not a 90° triangle, limiting options to sine law, or cosine law.

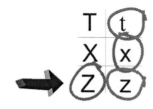

$$z^2 = 2^2 + 3^2 - 2(2)(3)\cos 60°$$

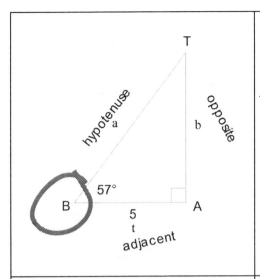

Side 'a'

Since this is a 90° triangle, select from the 3 basic ratios.

$$\cos 57° = \frac{5}{a}$$

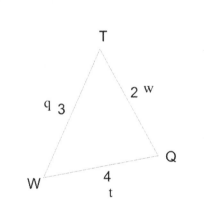

Smallest angle

This triangle is not a 90° triangle, limiting options to sine law, or cosine law. The smallest angle is opposite the shortest side.

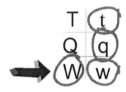

$$2^2 = 3^2 + 4^2 - 2(3)(4)\cos W$$

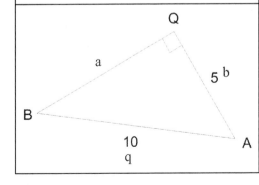

Side 'a'

Since this is a 90° triangle with 2 sides given, the 3rd side can be calculated using Pythagoras' Theorem.

$$a^2 + 5^2 = 10^2$$

DIAGRAM CONSTRUCTION

Each of the following equations was generated from a triangle, using some side(s), and some angle(s). Construct a triangle to illustrate a possible source for the equation. The triangles do not need to be drawn to scale.

$\dfrac{a}{\sin 25°} = \dfrac{32}{\sin 32°}$	
$b^2 = 5^2 + 7^2 - 2(5)(7)\cos 48°$	
$\cos C = 0.179$	
$\dfrac{\sin D}{5} = \dfrac{\sin 40°}{30}$	
$y^2 = 10^2 - 7^2$	

DIAGRAM SOLUTIONS

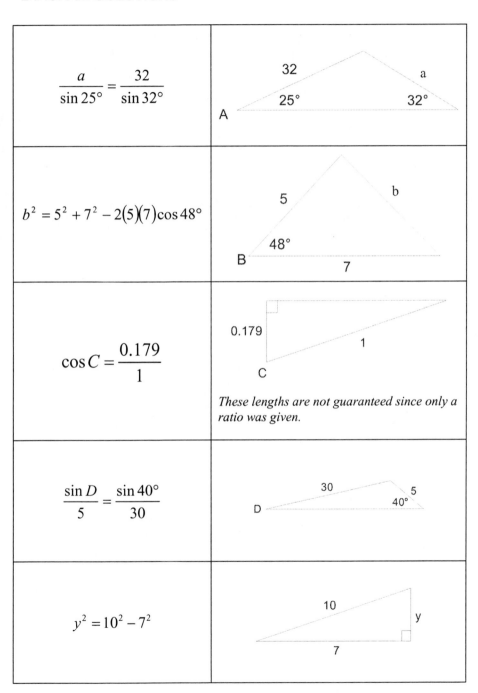

$$\frac{a}{\sin 25°} = \frac{32}{\sin 32°}$$

$$b^2 = 5^2 + 7^2 - 2(5)(7)\cos 48°$$

$$\cos C = \frac{0.179}{1}$$

These lengths are not guaranteed since only a ratio was given.

$$\frac{\sin D}{5} = \frac{\sin 40°}{30}$$

$$y^2 = 10^2 - 7^2$$

ERROR DETECTION

The following equations were generated from a set of triangles. Some may have one or more errors. Identify any error(s), and suggest a correct formula.

$\sin A = \dfrac{5}{4}$	
$b^2 = 3^2 + 5^2 - (3)(5)\sin 40°$	
$\dfrac{c}{\tan 25°} = \dfrac{10}{\tan 35°}$	
$\dfrac{d}{\sin 100°} = \dfrac{25}{\sin 85°}$	
$\tan E = 10$	

ERROR DETECTION SOLUTIONS

$\sin A = \dfrac{5}{4}$	*The sine ratio is defined* $\dfrac{opp}{hyp}$. *In this case the opposite side is longer than the hypotenuse. Perhaps the ratio was written upside down. Consider:* $$\sin A = \frac{4}{5}$$
$b^2 = 3^2 + 5^2 - (3)(5)\sin 40°$	*The formula is incorrect since the '2' in the equation was not written. Consider:* $$b^2 = 3^2 + 5^2 - 2(3)(5)\sin 40°$$
$\dfrac{c}{\tan 25°} = \dfrac{10}{\tan 35°}$	*There is no ratio formula for tangent. Consider:* $$\frac{c}{\sin 25°} = \frac{10}{\sin 35°}$$
$\dfrac{d}{\sin 100°} = \dfrac{25}{\sin 85°}$	*The measure of the angles in this formula total more than 180°. Perhaps the 100° was outside one of the triangle angles. Consider:* $$\frac{d}{\sin 80°} = \frac{25}{\sin 85°}.$$
$\tan E = 10$	*Although the sine and cosine ratio cannot be larger than 1, this restriction does not apply to the tangent ratio. This equation requires no correction.*

164

TRIGONOMETRY TEST

PART A   [10 marks]

Determine the numeric value of the variable in the following equations.

Lengths are required accurate to 2 decimal places.
The angle measures are required accurate to 1 decimal place.

1. $\sin(50°) = \dfrac{a}{30}$	1.
2. $\cos(B) = \dfrac{20}{35}$	2.
3. $\tan(60°) = \dfrac{5}{c}$	3.
4. $d^2 + 4^2 = 5^2$	4.
5. $e^2 = 10^2 + 5^2 - 2(10)(5)\cos(75°)$	5.
6. $\dfrac{f}{\sin(50°)} = \dfrac{5}{\sin(45°)}$	6.
7. $\dfrac{\sin(G)}{7} = \dfrac{\sin(25°)}{5}$	7.
8. $\cos(H) = \dfrac{1}{2}$	8.
9. $r = (\sin 45°)^2 + (\cos 45°)^2$	9.
10. $\tan J = 2.45$	10.

PART B   [5 marks]

<u>State the equation</u> that will solve for the variable stated in the following diagrams. There are no marks for solving the equations.

1. Determine the length of side 'q'.

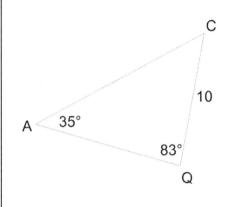

2. Determine the length of side 'k'.

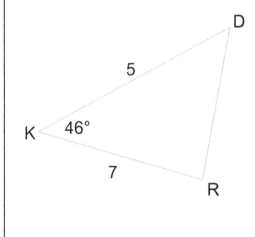

3. Determine the measure of angle '*B*'.

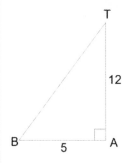

4. Determine the measure of angle '*R*'.

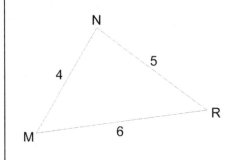

5. Determine the length of side '*a*'.

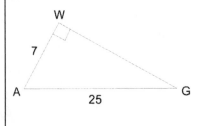

PART C   [25 marks]

Determine the length of the required side, or the measure of the required angle in the following diagrams. There is one mark for the answer. The remaining marks are determined by the quality of your solution.

1. Determine the length of side '*q*'. [4 marks]

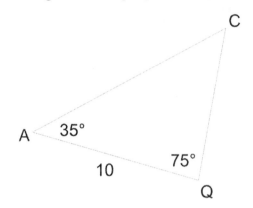

2. Determine the measure of angle '*P*'.   [4 marks]

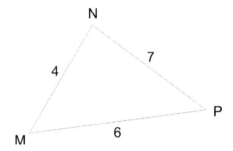

3. Determine the measure of the largest angle.   [5 marks]

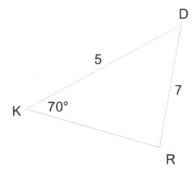

4. Determine the height '*BT*' in the following diagram. [5 marks]

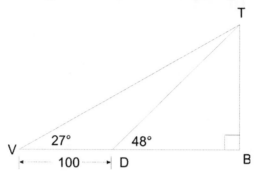

5. Provide a diagram that will allow someone to determine the height of the CN tower. All required information is to be clearly indicated. [3 marks]

6. Explain why the measure of angle 'C' in the equation $\sin C = \dfrac{5}{4}$ cannot be determined.   [2 marks]

7. State the one piece of information required in a triangle to be able to use any of the trig ratios, the sine law, or the cosine law.
    [1 mark]

BONUS:
Determine the length of side AB in the following triangle given the length of BD is 6 units, CQ is 8 units, and AC is 10 units.
[2 marks]

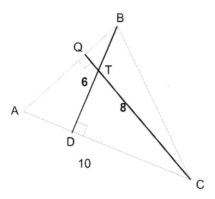

TRIGONOMETRY TEST SOLUTIONS

PART A SOLUTIONS [10 marks]

Determine the numeric value of the variable in the following equations.

Lengths are required accurate to 2 decimal places.
The angle measures are required accurate to 1 decimal place.

QUESTION 1	QUESTION 2
$\sin(50°) = \dfrac{a}{30}$ $a = 30\sin(50°)$ $a = 22.98$ *Less than 30 as expected, since the hypotenuse length is 30.*	$\cos(B) = \dfrac{20}{35}$ $B = \cos^{-1}\left(\dfrac{20}{35}\right)$ $B \approx 55.2°$
QUESTION 3	QUESTION 4
$\tan(60°) = \dfrac{5}{c}$ $c = \dfrac{5}{\tan(60°)}$ $c \approx 2.89$	$d^2 + 4^2 = 5^2$ $d^2 + 16 = 25$ $d^2 = 25 - 16$ $d^2 = 9$ $d = 3$ *Alternate: This is the 3, 4, 5 triangle.*

QUESTION 5	QUESTION 6
$e^2 = 10^2 + 5^2 - 2(10)(5)\cos(75°)$ $e^2 = 125 - 100\cos(75°)$ $e^2 \approx 125 - 25.8819$ $e^2 \approx 99.1181$ $c = \sqrt{99.1181}$ $e \approx 9.96$ *This side length is between 5, and 15, the two possible lengths for the extreme triangles with an angle measure of 0°, or 180° between the 2 given sides.*	$\dfrac{f}{\sin(50°)} = \dfrac{5}{\sin(45°)}$ $f = \dfrac{5\sin(50°)}{\sin(45°)}$ $f \approx 5.42$ *Longer than the 5 with the 45° opposite it, since side 'f' is opposite a larger angle.*
QUESTION 7	QUESTION 8
$\dfrac{\sin(G)}{7} = \dfrac{\sin(25°)}{5}$ $\sin(G) = \dfrac{7\sin(25°)}{5}$ $\sin(G) \approx 0.5917$ $G = \sin^{-1}(0.5917)$ $G \approx 36.3°$ or $143.7°$ *The second angle is calculated by subtracting from 180°.* *This angle can be checked by typing $\sin 143.7°$ expecting approximately 0.592 as the answer.*	$\cos(H) = \dfrac{1}{2}$ $H = \cos^{-1}\left(\dfrac{1}{2}\right)$ $H = 60.0°$
QUESTION 9	QUESTION 10
$r = (\sin 45°)^2 + (\cos 45°)^2$ $r = 0.5 + 0.5$ $r = 1.0$	$\tan J = 2.45$ $J = \tan^{-1}(2.45)$ $J \approx 67.8°$

PART B [5 marks]

---

1. Determine the length of side '*q*'.

This triangle is not a 90° triangle, limiting options to sine law, or cosine law.

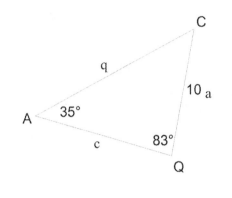

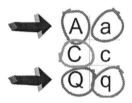

$$\frac{q}{\sin(83°)} = \frac{10}{\sin(35°)}$$

---

2. Determine the length of side '*k*'.

This triangle is not a 90° triangle, limiting options to sine law, or cosine law.

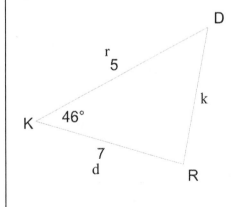

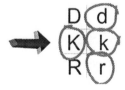

$$k^2 = 5^2 + 7^2 - 2(5)(7)\cos(46°)$$

3. Determine the measure of angle 'B'.

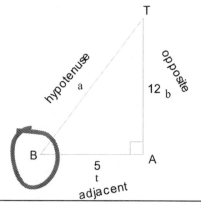

The triangle has a 90° angle. Use one of the basic trig ratios.

$$\tan B = \frac{12}{5}$$

4. Determine the measure of angle 'R'.

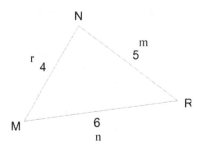

This triangle is not a 90° triangle, limiting options to sine law, or cosine law.

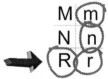

$$4^2 = 5^2 + 6^2 - 2(5)(6)\cos R$$

5. Determine the length of side 'a'.

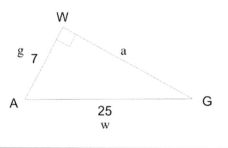

Since the triangle has a 90° angle, and two sides are given, Pythagoras' Theorem is the best method. The hypotenuse length is 25.

$$a^2 + 7^2 = 25^2$$

PART C   [25 marks]

Determine the required side, or angle in the following diagrams. There is one mark for the answer. The remaining mark is determined by the quality of your solution.

1. Determine the length of side 'q'. [4 marks]

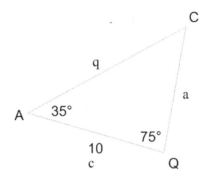

This triangle is not a 90° triangle, limiting options to sine law, or cosine law.

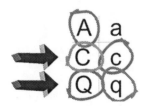

Angle C is available
$180° - 35° - 75° = 70°$.

$$\frac{q}{\sin 75°} = \frac{10}{\sin 70°}$$

$$q = \frac{10\sin 75°}{\sin 70°}$$

$$q \approx 10.28$$

The length of side 'q' is expected longer than 10 since the opposite angle to side 'q' is larger than the 70° angle opposite the side length of 10.

2. Determine the measure of angle 'P'.   [4 marks]

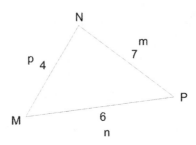

This triangle is not a 90° triangle, limiting options to sine law, or cosine law.

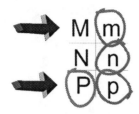

$$4^2 = 7^2 + 6^2 - 2(7)(6)\cos P$$
$$16 = 49 + 36 - 84\cos P$$
$$16 = 85 - 84\cos P$$
$$84\cos P = 85 - 16$$
$$84\cos P = 69$$
$$\cos P = \frac{69}{84}$$
$$P = \cos^{-1}\left(\frac{69}{84}\right)$$
$$P \approx 34.8°$$

Watch the order of operations.

Angle P is acute since the ratio is positive.

Only one angle is possible, but if $\sin^{-1}$ were involved, there would be a second angle to consider.

3. Determine the measure of the largest angle.   [5 marks]

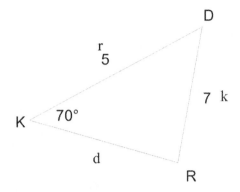

D

r
5

7 k

K   70°

d

R

This triangle is not a 90° triangle, limiting options to sine law, or cosine law.

Angle 'R' will be calculated 1ˢᵗ, and then angle 'D' will be available.

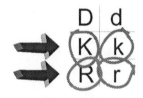

$$\frac{\sin R}{5} = \frac{\sin 70°}{7}$$

$$\sin R = \frac{5\sin 70°}{7}$$

$$\sin R \approx 0.6712$$

$$R \approx 42.2° \text{ or } 137.8°$$

$$R \approx 42.2°$$

Keep all decimals in the calculator memory.

The 137.8° angle is not possible since together with the 70° angle, the measure of the angles in the triangle will total more than 180°.

The 3rd angle will be 180° - 70° - 42.2° = 67.8°, making angle 'K' the largest, at 70°.

4. Determine the height '*BT*' in the following diagram.
    [5 marks]

There are 3 triangles available. This solution will use the 2 enclosed triangles VTD, and DTB. First, calculate the other angles.

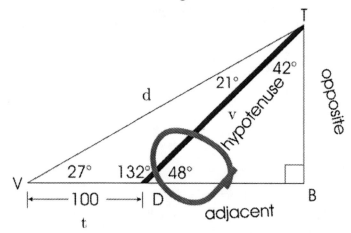

Using side DT for both smaller triangles:

USING VTD	**USING DTB**
(LEFT TRIANGLE)	(RIGHT TRIANGLE)

This triangle is not a 90° triangle, limiting options to sine law, or cosine law.

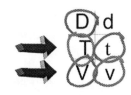

$$\frac{DT}{\sin 27°} = \frac{100}{\sin 21°}$$

$$DT = \frac{100\sin 27°}{\sin 21°}$$

$$DT \approx 126.68$$

$$\sin 48° = \frac{BT}{126.68}$$

$$BT \approx 126.68\sin 48°$$

$$BT \approx 94.14$$

Remember to keep all decimals until the final answer (if possible).

ALTERNATE SOLUTION

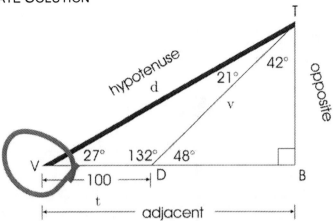

Using side VT from the left triangle, and the large triangle:

USING VTD	USING VTB
(LEFT TRIANGLE)	(LARGE TRIANGLE)

This triangle is not a 90° triangle, limiting options to sine law, or cosine law.

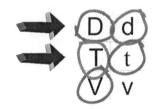

$$\frac{VT}{\sin 132°} = \frac{100}{\sin 21°}$$

$$VT = \frac{100\sin 132°}{\sin 21°}$$

$$VT \approx 207.37$$

$$\sin 27° = \frac{BT}{207.37}$$

$$BT \approx 207.37 \sin 27°$$

$$BT \approx 94.14$$

Remember to keep all decimals until the final answer (if possible).

Both methods have the same final answer. The height BT is approximately 94 units in length.

5. Provide a diagram that will allow someone to determine the height
    of the CN tower. All required information is to be clearly
    indicated. [3 marks]

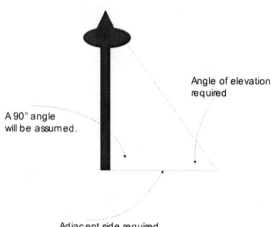

*Possibly, the angle of elevation will be measured from eye height. If that is
the case, add this eye height to the initial calculated height of the tower.*

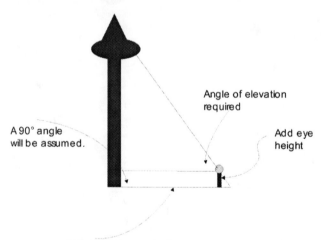

*Note: the top of the tower moves with the wind, possibly requiring a day
for measuring when there is no wind.*

6. Explain why the measure of angle 'C' in the equation $\sin C = \dfrac{5}{4}$ cannot be determined.   [2 marks]

The definition of the sine ratio is $\sin B = \dfrac{opposite}{hypotenuse}$ *, with the hypotenuse as the longest side in a 90° triangle. The ratio in the equation then, is not possible since the opposite side is stated larger than the hypotenuse. The calculator will recognize the sine ratio will never be greater than 1.*

7. State the <u>one piece of information required</u> in a triangle to be able to use any of the trig ratios, the *sine law*, or the *cosine law*. [1 mark]

All triangles must have <u>at least one side length</u> available for any calculations to be completed. Without a length, there is no unique triangle size possible. Using the trig summary on page 114, more information will still be required.

BONUS:

Determine the length of side *AB* in the following triangle given the length of *BD* is 6 units, *CQ* is 8 units, and *AC* is 10 units.
[2 marks]

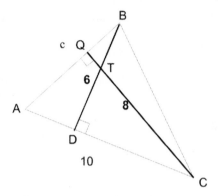

The formula for area of a triangle is

$$A = \frac{1}{2}bh$$
.

Using this formula twice, each time with a different base, the length AB can be determined without using any trig formula.

Left Side uses AC as the base.
Right Side uses AB as the base (letter 'c').

$$\frac{1}{2}(10)(6) = \frac{1}{2}(c)(8)$$

$$30 = 4c$$

$$4c = 30$$

$$c = 7.5$$

COSINE LAW
DEVELOPED

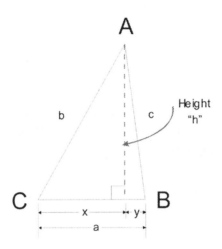

Everything we have used with the three trig ratios, and Pythagoras require a 90°, or 'right' triangle. Since this triangle has no right angle, a height 'h', will be introduced *temporarily*. This height divides the base into two temporary sections, 'x', and 'y'. Since the original triangle has three angles A, B, and C with matching opposite sides a, b, c, these letters will be acceptable in the final formula. Any extra variables that are introduced will be treated as temporary.

The two 90° triangles allow a Pythagorean equation for each, and the divided side allows a third equation.

From the left triangle	From the right triangle	From the base shown
$[1] \quad b^2 = x^2 + h^2$	$[2] \quad c^2 = y^2 + h^2$	$[3] \quad x + y = a$

Since 'h' was introduced temporarily, is should be eliminated as soon as possible. From equation [1] $h^2 = b^2 - x^2$. Now substitute into equation [2]. The following occurs:

$$[2] \quad c^2 = y^2 + h^2$$
$$\textit{from } [1] \quad c^2 = y^2 + \left(b^2 - x^2\right)$$
$$\textit{simplified} \quad c^2 = b^2 + y^2 - x^2$$

This equation has 2 triangle variables, 'b', and 'c'.
Using equation [3], decide to solve for 'y'. This gives: $y = a - x$.
Now substitute this into the latest equation. The following occurs:

$$c^2 = b^2 + y^2 - x^2$$
$$c^2 = b^2 + \left(a - x\right)^2 - x^2$$
$$c^2 = b^2 + \left(a - x\right)\left(a - x\right) - x^2$$
$$c^2 = b^2 + a^2 - ax - ax + x^2 - x^2$$
$$c^2 = b^2 + a^2 - 2ax$$

This equation has 3 triangle variables, 'a', 'b', and 'c'. An extra letter 'x', also was introduced as part of side 'a'. Using the left side right triangle, with 'x' as the adjacent side, and the hypotenuse 'b', we can introduce one more equation.

$$\cos C = \frac{x}{b}$$
$$\frac{\cos C}{1} = \frac{x}{b}$$
$$x = b \cos C$$

S O H
C A H
T O A

The final substitution produces the *cosine law*.

$$c^2 = b^2 + a^2 - 2ax$$
$$c^2 = b^2 + a^2 - 2a(b\cos C)$$
$$c^2 = b^2 + a^2 - 2ab(\cos C)$$
$$c^2 = a^2 + b^2 - 2ab(\cos C)$$

With the order rearranged.

By changing the labels on the triangle, the other 2 cosine formulas for this triangle can be easily obtained.

$$a^2 = b^2 + c^2 - 2bc(\cos A)$$
$$b^2 = a^2 + c^2 - 2ac(\cos B)$$

Printed in the United States
108228LV00004B/361/A